W9-DJK-476

301.3608
R49

108655

DATE DUE			

WITHDRAWN

INTERNATIONAL EDITORIAL ADVISORY BOARD

ROBERT R. ALFORD
University of California, Santa Cruz

HOWARD S. BECKER
Northwestern University

BRIAN J.L. BERRY
Harvard University

ASA BRIGGS
Worcester College, Oxford University

JOHN W. DYCKMAN
University of Southern California

H.J. DYOS
University of Leicester

T.J. DENIS FAIR
University of Witwatersrand

SPERIDIAO FAISSOL
Brazilian Institute of Geography

JEAN GOTTMANN
Oxford University

SCOTT GREER
University of Wisconsin, Milwaukee

BERTRAM M. GROSS
Hunter College, City University of New York

ROBERT J. HAVIGHURST
University of Chicago

EIICHI ISOMURA
Toyo University

ELISABETH LICHTENBERGER
University of Vienna

M.I. LOGAN
Monash University

WILLIAM C. LORING
Center for Disease Control, Atlanta

AKIN L. MABOGUNJE
University of Ibadan

MARTIN MEYERSON
University of Pennsylvania

EDUARDO NEIRA-ALVA
CEPAL, Mexico City

ELINOR OSTROM
Indiana University

HARVEY S. PERLOFF
University of California, Los Angeles

P.J.O. SELF
London School of Economics and Political Science

WILBUR R. THOMPSON
Wayne State University and
Northwestern University

THE RISE OF THE SUNBELT CITIES

Edited by
DAVID C. PERRY and
ALFRED J. WATKINS

CARL A. RUDISILL LIBRARY
LENOIR RHYNE COLLEGE

Volume 14, URBAN AFFAIRS ANNUAL REVIEWS

SAGE PUBLICATIONS / BEVERLY HILLS / LONDON

301.3608
R49
108655
Feb. 1979

Copyright © 1977 by Sage Publications, Inc.

All rights reserved. No part of this book may be reproduced or utilized in any form or by any means, electronic or mechanical, including photocopying, recording, or by any information storage and retrieval system, without permission in writing from the publisher.

For information address:

SAGE PUBLICATIONS, INC.
275 South Beverly Drive
Beverly Hills, California 90212

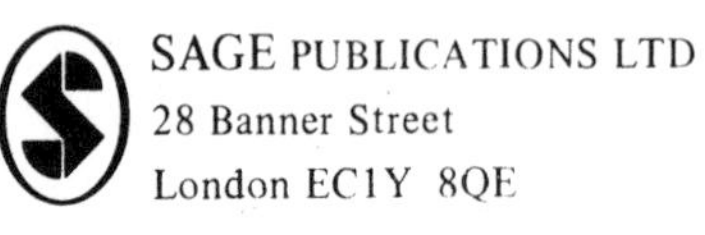

SAGE PUBLICATIONS LTD
28 Banner Street
London EC1Y 8QE

Printed in the United States of America

International Standard Book Number 0-8039-1029-0 (cloth)
0-8039-1030-4 (paper)

Library of Congress Catalog Card No. 77-93698

FIRST PRINTING

CONTENTS

Introduction

DAVID C. PERRY
ALFRED J. WATKINS

□ THIS BOOK IS CONSCIOUSLY ORGANIZED to critically address and refine a series of the fundamental disciplinary elements comprising American urban studies. The goals of the contributors have been larger than simply shedding more analytic light on the issues surrounding the present state of regional tension in the United States. We have all attempted to supply new empirical insights within new boundaries of historical and theoretical definitions of urbanism in America. As such, this volume hopefully represents not only some measure of advancement of our knowledge about urban development but also a useful contribution to the intellectual expansion of the discipline of urban studies. The extent to which these highly ambitious goals have been achieved is directly attributable to the eight month process of interaction between the editors and the authors. Every contributor spent untold numbers of hours with the editors in a supreme effort to make this project work.

THE GENERAL GOALS OF THIS BOOK

In a very real sense this book represents a departure from the traditional foci of literature on the contemporary state of the American city. First, its topic is the growth and development of the

Southern and Southwestern cities of the United States. As such, it represents a departure from the most distinct empirical bias in the urban literature: namely, the overall dependence of urban scholars upon the character of older Northeastern cities as the definitive frame through which all cities in the U.S. are analyzed. The Northeastern metropolis with its relatively small central city and its expansive and highly populated suburbs has acted as an urban paradigm for everything from the U.S. Bureau of the Census's definition of a Standard Metropolitan Statistical Area to the assumption that all urban dynamics could be characterized in an intrametropolitan fashion. Thus, the successes and failures of urbanization could be captured by splitting metropolitan areas into sinking central cities floundering in a sea of sprawling suburbs. Adherents and critics alike found more than enough substance in the material decline and rising social decay of the central city and the economic growth and materialistic excesses of the suburbs of the northeastern urban centers to power their assumptions that such conditions accurately reflected the dominant motif of all but a few urban centers.

In part, this explains the level of surprise registered by both the national press and the academic world as the national economic order apparently shifted on its head with the "rise of the Sunbelt." That people and industries would leave New York, Chicago, and Boston to live in the "backwoods" of Georgia, with the "cowboys" of Texas, or in the "deserts" of Arizona was simply unthinkable. While these states were part of America they did not house our *urban* centers. Even more disconcerting was the fact that Sunbelt SMSAs quite often did not at all approximate the northeastern model. Their central cities housed the lion's share of their metropolitan populations, economic bases, and "suburban" areas. Their tax bases were not shrinking and their public service levels were not extreme. In short, for analysts, they represented a culturally alien and "non-urban" set of urban areas.

Such myths surrounding the regions of the Sunbelt were quite understandably fueled by a centuries-old pattern of uneven economic development. Given this perspective of interregional development, it was not surprising that scholars would find the historically less powerful urban centers of the South and Southwest less legitimate objects for study. The Northeast had been the dominant regional center for the most productive activities of our national economy and the "Yankee" cities had emerged as the pivotal axes around which our economy swirled. The South and the Sunbelt were resource-rich "outlands" and their cities were the subordinate subcenters of our emerging economy.

However, it appears as if much of this past tradition has been profoundly altered. Now, academics, the press, and politicians supply us with a new description of urban America. They cite a regional "power shift," inaugurate Houston as the "new diamond studded buckle" of America's economic empire, and claim that New York City and other Northeastern cities may no longer just be "decaying"—they may be actually dying. However dramatic and surprising these trends *appear,* it is important to emphasize that the *Rise of the Sunbelt Cities* did not occur overnight.

The authors of the essays in this book take great care to emphasize this point. Their essays, focusing as they do on Southern and Southwestern cities, represent a direct critique of the sociology of urban knowledge generated by the biases suggested above. The history of the socioeconomic development of cities in the Sunbelt has indeed been different, in many important respects, from the development of Northeastern cities, but such differences in development were not cause for the dearth of study and the concomitant misspecification of Sunbelt cities which is found in most contemporary urban studies. As such, these essays start urban scholarship on the road to a more balanced empirical vision. They offer detailed analyses of urban centers which have represented regionally specific and nationally important chapters in the history of our urban people.

Further, this book represents a realistic departure from the overdependence of many contemporary scholars on an intrametropolitan perspective to urban analysis. Just as regional economies are centered in cities so, too, are cities acting in environments that far exceed their metropolitan hinterlands. For too long, the dominant scholarship of the urban social sciences has paid only lip service to the importance of regional economies and cultures, relegating such topics to the fringes of the disciplines. The essays in this volume represent a studied effort to direct our attention to the intermetropolitan and interregional as well as the intrametropolitan character of American urbanization. Cities are not independent communal actors; they are influenced by their immediate surroundings and by the particular regional, national, and even international urban constellations of which they are a part.

Finally, the essays in this volume, hopefully, represent a measured response to the growing gale of rhetoric building around the "second war between the states." The intent of all the authors has been to supply a new round of observation and analysis to the politically and economically shifting currents presently at work in urban America. The pieces attack with rigor the rhetoric accompanying these shifts and supply often competing, but seriously reasoned, discussions of the changing nature of Northeastern and Sunbelt cities. As such this volume represents an attempt to explain contemporary urban change and pragmatically rechannel or shut off some of the more popular streams of short sighted analysis.

Therefore, in light of these trends, we present a series of new directions for urban scholarship. The three essays of Part I supply a mixture of traditional, neoclassical, and Marxist explanations of the developmental history of Northeastern and Sunbelt cities. We believe that all of these pieces chart substantially new theoretical approaches to the past and present conditions of urban growth and decay. While they come from somewhat competing directions, the authors set a comprehensive stage for not only understanding contemporary urban conditions but also the state of urban scholarship.

The second section includes six essays on themes which have always been central to the growth and development of American cities: the role of the private sector; the role of the media; the nature of urban political culture; and the role of the public sector. What differentiates these essays from other works is that these essays represent specific attempts to study these important aspects of urban society in the rarely considered frame of Southern and Southwestern cities. Each of the pieces critically assails the popular and scholarly myths concerning the practices of business, government, media, and culture in the urban Sunbelt while supplying useful new case materials and conceptual insights.

In Part III, the writers offer three distinct perspectives on the future of American cities. The assessments are frank and urge a drastic reappraisal of the urban conditions in America. At the most particular level, they all emphasize the profound dangers inherent in continued acquiescence to the trends found in the present round of urban development. More generally, two of the essays critically attack the primary role the urban unit has played in the evolution of urban society, arguing that nothing short of a redefinition of the urban function will be sufficient to end the "crises" which characterize such phenomena as "the rise of the Sunbelt cities" and the "decline of the Northeast."

A NOTE ON THE WRITING OF THIS BOOK

One further component of this book must be discussed here. The authors represent a wide range of academic and political spectra. However, the actual process of preparing this book has been neither a sterile exercise in the "multidisciplinary" tradition of the academy nor has it evolved into a deafeningly rhetorical debate over the competing implications of our analyses. Each of the selections is the result of countless hours of discussion, worry, scotch tape, erasers, pain, and friendship. It is difficult to record the level of personal interest and the amount of time the authors shared with the editors on every phase of this project. We started out to edit a book on the Sunbelt cities and were treated to the finest eight-month seminar on the topic two people could ever attend.

In addition to the authors and editors there is another group of people who contributed directly to the completion of this volume. The substantive and informative support we have received from Joe Feagin, Jose Angel Guttierez, Bobby Mangiani, K.C. Cerny, John Mollenkopf, Karl Schmitt, Jeff Millstone, and Donald Brittain, our editor at Sage, is profoundly appreciated. We are especially grateful to Steve Pasternack for the countless hours he has spent with us on this project. We could very easily include Sara Norwood as the third editor of this volume; she has been with us every "word of the way" as researcher, critic, and colleague. The task of writing this book would have been immeasurably more difficult without the financial support of the Earhart Foundation and Dean Robert King of the School of Social and Behavioral Sciences at the University of Texas at Austin.

Above all, in the process we have learned as much about each other as we have about the Sunbelt. We are friends. This book has been a chance for a great number of people, from the Northeastern Industrial Rim and the Sunbelt, to learn from each other. In the process, we made a book. The successes go to the authors, the failures go to the editors, and the first draft of this introduction was flushed down the toilet by a two year old named Evan Perry. A second critic, his older brother Clayton, says the 14th volume of the *Urban Affairs Annual Reviews* needs a few pictures. Maybe it does, but it's hard to find art in the writing of urban scholars. We hope at least the reader finds an insight or two.

Part I

Three Theories of American Urban Development

Introduction

□ ANYONE WHO HAS PAID CLOSE ATTENTION to the daily outpouring of news items emanating from the American journalistic establishment cannot help but to have noticed a dramatic change in the coverage of urban affairs. During the 1960s, attention was riveted to such topics as police brutality, riots in the streets, law and order, urban renewal, and the flight to the suburbs. That decade's urban crises were expressed in terms of intrametropolitan dualities—city versus suburb, black versus white, ghetto versus everyone else. Although it was painfully evident that the central cities, and especially certain population groups and neighborhoods, were in serious economic trouble, a strange complacency dominated the rhetoric of the era. Despite their fall from economic grace, these cities were surrounded by healthy, prosperous suburban communities and if we could only channel some of this peripheral wealth back to the center of the metropolitan region, all would be well in the best of all possible worlds. Based on this paradigmatic analysis, policies were formulated to alter the housing patterns of disadvantaged central city residents, and studies were designed to examine the feasibility of restructuring the urban transportation network so that people could travel to jobs in the prosperous sections of the metropolitan community. The concept of new towns in town was promulgated in an attempt to stem the white flight which was enervating the fiscal vitality of these core areas. The Great Society was inaugurated to upgrade the job skills and formal training of the ghetto residents so that they could benefit from the economic boom that was spreading out all around them.

But programs of well-intentioned prescriptions do not guarantee the desired results and by the late 1960s the optimism that infused urban research during the middle of the decade was shattered. No longer did people believe that a minor dose of social tinkering would be sufficient to restore the economic vitality of the central cities. The problems were too profound and deep seated to be solved by the available policies. Thus, pessimism replaced optimism and hope was transformed into skepticism. Despite the efforts of government officials, the decline in many of the older metropolitan areas was proceeding at an alarming rate and, now, the symptoms of decay were making their first appearance in the suburbs. The *cordon sanitaire* was breached and no segment of the metropolitan community could safely assume that blight would not invade their sanctuary. Then came New York City's flirtation with bankruptcy and financial chaos. The old palliatives of welfare programs designed to assuage the demands of ghetto residents could be applied no longer and the rising tide of joblessness could not be cosmetically covered by increasing the employment opportunities in the local government sector.

In the midst of these crumbling illusions, a new variable was injected into the urban crisis scenario. Not only were the old central cities of the Northeast sick and decaying, but entire metropolitan areas were losing population and jobs to a younger and more vigorous section of the country. The "Cowboys" were vanquishing the "Yankees" in a "second war between the states" that was permanently transforming the regional landscape. As the cities of the Northeast, buried chin deep in snow, were declining, a new urban frontier was opening in the Sunbelt fueled by low wages, a relatively unfettered arena for capital accumulation, and sunshine.

Both journalists and academicians were caught by surprise at the seeming rapidity of this economic transformation. In many respects this lack of awareness is inexcusable though perhaps understandable in view of the prevailing state of urban analysis. The rise of the Sunbelt was not an overnight phenomenon. Its antecedents were deeply rooted in past historical events but, in most cases, the shifting rhythms and patterns of historical evolution were never incorporated into discussions of urban affairs. Instead, an emphasis on empirical techniques designed to describe the current state of affairs replaced most attempts to understand these events through theory construction. This produced a situation of disciplinary isolation from real world phenomena. Social scientists were surprised by the obvious because the dominant techniques left them with a paucity of both hindsight and vision. They could not see where the country was going because they did not understand where it had been.

In this section, we have tried to overcome these defects and set a new tone for urban scholarship which will change the nature of the dialogue about the American urban condition. Each of the three selections, despite both their different theoretical and philosophical underpinnings and their emphasis on different causal variables, share several surprising areas of agreement. First, and perhaps foremost, they all agree that the rise of the Sunbelt is not some unexpected anomaly in the history of U.S. urban development; rather it is part of a long-standing and ongoing process which has continuously transformed the nation's system of cities. By relying on historical analysis, each of the authors attempts to illuminate one important facet of this process. Second, all three selections argue quite forcefully that urban development cannot be understood in an institutional vacuum. Within the U.S. context, the authors agree, the primary variable is not transportation, technology, or some of the other factors which have received such prominent attention in past analyses. In their opinion, the evolution of the capitalist system creates a changing developmental and regional growth environment. For various reasons, some cities and regions are better situated to take advantage of these changes. These soon become the most dynamic metropolitan centers while their less fortunate counterparts are left to languish with slow growth and mounting social problems. Finally, all of the selections suggest that despite the evolutionary nature of social and economic change, several fairly distinct historical epochs can be delineated. Within each period, different dynamic processes dominate the economic landscape and, as one epoch is transformed into the next, this has profound effects on the relative prosperity and rates of growth in different metropolitan areas.

The first selection by Alfred J. Watkins and David C. Perry suggests that the rise of the Sunbelt represents one more episode in a long-standing process of uneven urban development. In their opinion, each stage of capital accumulation is strongly correlated with a new phase of urban development. New cities emerge by capturing a significant portion of the new dynamic activities spawned within each stage while the older cities remain locked into old economic activities, thus limiting their ability to capitalize on the requirements of the new epoch. The authors suggest that the phenomenon of uneven development results from a shifting array of barriers and opportunities generated by each epoch. Specifically, those cities which are rapidly ascending the urban hierarchy protect their dominant position by establishing a set of institutional and structural barriers which limit the number of competing centers. As a result, the subordinate metropolitan areas serve as colonial appendages, producing wealth which primarily benefits the dominant urban areas. However, as the existing era wanes, the addiction to the old methods and practices becomes a barrier blocking further growth in the dominant centers. The formerly subordinate cities, by dint of their peripheral position in the old epoch and their lack of attachment to the old economic activities, are perfectly situated to capitalize on the opportunities generated by the new wave of capital accumulation. In this way, the mantle of growth is continuously shifted from one region to the next and the result is a pattern of uneven regional development.

David M. Gordon's piece explicitly rejects the notion that technological change provides the primary impetus for regional and urban development. Terming such explanations "technological determinism," Gordon suggests that labor discipline is of paramount importance. According to Gordon's conception, technology must be evaluated on two distinct planes. On the one hand, a specific method of production can be termed efficient if it produces the maximum output while minimizing the required inputs. On the other hand, however, a given technology will be socially efficient only if it succeeds in maintaining the system defining class structure with its hierarchical relations of dominance and subordinance. By explaining the interaction of these two dimensions, Gordon accounts for the internal transformations within cities as well as the regional realignment of economic power.

Walt W. Rostow's article attributes regional dynamics to the relative price movements which alternately favor one region at the expense of the other. Drawing upon the statistical studies of N.D. Kondrateiff, who analyzed the historical fluctuations of wages, prices, and interest rates, Rostow suggests that we are in the midst of a fifth Kondratieff upswing. During each previous upswing, a rise in the relative prices of basic commodities has occurred and it has been accompanied by a general inflation, high interest rates, pressure on real wages, and most importantly, shifts of income favorable to producers of food and energy. In view of these trends, Rostow suggests that the decline in the Northeast can be explained, at least in part, by their unfavorable terms of trade. At the same time, these price movements are favoring the Sunbelt which specializes in producing precisely those commodities which are in short supply and whose prices are rising. Unless and until steps are taken to correct this imbalance, Rostow forsees no changes in the basic pattern of regional dynamics.

1

Regional Change and the Impact of Uneven Urban Development

ALFRED J. WATKINS
DAVID C. PERRY

□ THE EMERGENCE OF THE SUNBELT CITIES as the new frontier of dynamic urban growth has recently captured the attention of social scientists. But while the demographic and economic changes accompanying this "power shift" have been fairly well documented (Sale, 1975; Sternlieb and Hughes, 1975), the attempts to situate this phenomenon within the broader context of U.S. economic development and the evolution of the urban system have spawned a variety of confusing and contradictory explanations. In this chapter, we will critically reassess the most popular of these theories—that of economic convergence—and then supply what we believe to be a more comprehensive and penetrating historical approach. We will suggest that the apparent shift in power from the "Yankee" to the "Cowboy" urban areas is just one more episode in a continuing and long-standing process of uneven urban development.

THE PREVAILING EXPLANATION OF ECONOMIC CONVERGENCE

In the opinion of the convergence theorists (Williamson, 1965; Borts and Stein, 1964), the Northeastern cities are not dying;

instead they are merely laboring under the handicaps imposed upon them by an economic system adjusting to a new spatial equilibrium. Spurred on by the greater productive efficiency emanating from cheap land, low wages, low taxes, and a "favorable business climate," and led by Houston—"the shining buckle of the sunbelt"—newer cities are now completing the inevitable process of convergence initiated before the turn of the century.

Generally speaking, the proponents of convergence argue that the Northeast's urban areas are not necessarily handicapped by some fatal, chronically debilitating illness. Instead their problems result from the fact that the mantle of growth has temporarily shifted to those metropolitan areas with a more favorable constellation of factor costs. However, once a state of equilibrium is reestablished between the two regions, the extreme growth disparities can be expected to level off and all cities will grow in a more harmonious and uniform fashion. Unfortunately, this equilibrating convergence process is not painless—the rich, more developed, region will be made slightly poorer while the poor and less developed region will become slightly richer. Thus according to these theorists, the present problems in the Northeast are not necessarily the harbingers of an alarming state of crisis. Rather they are only a reaffirmation of the existence of a self-regulating market mechanism.

Upon closer examination, however, it is not at all clear what is meant by convergence. What exactly is converging in the two regions? Does convergence imply that the two regions are developing homogeneous industrial structures or is it simply a statement that the levels of development within the two regions are becoming more equal? Finally, what is the process through which the poor region begins to overtake its more developed counterpart?

According to its most accepted usage, convergence implies an increasing uniformity in various macro-level indicators—i.e., per capita income, growth rates, factor costs, the sectoral distribution of the labor force, and so on. Most of the empirical evidence seems to lend at least some credence to these hypothesized trends (Kuznets et al., 1960). Regional differences in per capita income have been narrowed, cost factor differentials, while not totally eliminated, have been reduced, and those areas whose economic base had formerly been dominated by agricultural pursuits have shifted to a more balanced industrial foundation. If this is the meaning of convergence, there is little to quibble over. Also, on a sociocultural level, there is little doubt that regional peculiarities are moderating. Radio, television, and the cinema have all had a leavening effect. Within the Sunbelt, the shift of manufacturing plants from scattered rural locations to larger urban centers brought the tempo of life into greater accord with that of the more urbanized east. And finally, the general flow of population from rural to urban areas has meant that small town values have been diffused by the more cosmopolitan outlook of metropolitan America. Thus, on this level also, convergence is a meaningful concept.

But how does convergence occur? Certainly, the proponents of convergence do not wish to imply that Sunbelt cities are developing industrial structures which are carbon copies of Northeastern cities. Although income levels and occupational distributions are becoming more similar, it would be ludicrous to argue that Tampa, Phoenix, and Las Vegas are, in some qualitative sense, similar to Philadelphia, Newark, and Akron. Moreover, it is rather doubtful that the present political and economic leaders of Houston or Austin are advocating that their respective cities emulate some of the better known qualities of New York City, such as municipal bankruptcy, a shrinking economic base, power failures, and gutted neighborhoods. It is clear, therefore, that in certain contexts the notion of convergence is a relatively valid concept. Within other contexts, however, convergence is totally inapplicable. Thus, it behooves the proponents of convergence to define more precisely exactly what elements of the economic system are converging and to explain the process by which convergence does occur.

On a more theoretical plane, convergence theory is predicated upon the dual set of beliefs that there is (a) a persistent tendency for the economies of two regions to conform to the conditions associated with a long run general equilibrium and that (b) any forces which disturb the system will be eventually offset by equal and opposite counteracting pressures. However, this mechanistic view of the equilibration process has been questioned by Myrdal (1957), who rejects the notion of convergence and argues that regional adjustments can be better explained by using the concepts of cumulative causation and deviation amplifying feedback cycles. According to this scenario, "a system is by itself not moving toward any sort of balance between forces but is constantly on the move away from such a situation. In the normal case, a change does not call forth countervailing changes but, instead, supporting changes, which move the system in the same direction as the first change but much further" (Myrdal, 1957:13).

Although aware of Myrdal's formulation of the regional adjustment process, many of the treatises on convergence content themselves to merely mention it in passing before proceeding with a discussion of convergence. However, an equally valid regional development scenario can be outlined based upon cumulative causation and deviation amplifying feedback cycles. Within this framework, population and economic changes would be self-reinforcing. The local economy would gain momentum since a change in one variable would provide an incentive for ensuing changes of the same direction in the other variable. Thus, expanding local economies would continue to expand while depressed areas would face a continuous shrinkage in their industrial base. Divergence, not convergence, would ensue.

If this scenario is accepted as an accurate portrayal of urban dynamics, then the concept of long run stable equilibrium must be discarded. Cities on disparate growth trajectories are not headed toward some common point but, on the contrary, are moving further away from each other. Those studies which have attempted to empirically demonstrate the tendency for regional economies to converge over time are ignoring the essential dynamic qualities of economic evolution. Rather than convergence, a more apt analogy can be drawn by considering two trains headed in opposite directions. Up to a certain point, the trains are converging but not necessarily for the purpose of coming to rest once they have met. Instead, the period of convergence is only a necessary prelude to increasing divergence. The trains are still headed in opposite directions even though it appears that they are converging.

Studies which demonstrate that per capita incomes, industrial structure, and labor force participation rates are all converging toward the establishment of a relatively homogeneous system of cities are, therefore, guilty of two flaws. In the first place, they are only capturing the first half of the the process–that portion which is relevant until the two trains meet. Second, they are employing the technique of comparative statics. Statistics for two groups of cities are compared at two different points in time. The intervening period is ignored and, as a result, the paths of motion for each group of cities is never examined. Hence, the direction of the long run secular trends are wiped completely from consideration with the result that divergent aspects of interregional economic development are often misconstrued as representing convergence.

A THEORY OF UNEVEN URBAN DEVELOPMENT

If we replace the concept of general equilibrium with a scenario which places primacy upon cumulative causation, then the process of convergence becomes much more problematic. Uneven divergent development and not convergence would appear to be a more typical representation of interregional economic adjustments. In this section, we will introduce some of the salient theoretical arguments in support of uneven interregional development.

On an elementary level, the concept of uneven urban and regional development implies that capital accumulation and the location of rapidly growing industries are not evenly dispersed throughout the nation. Rather, they are concentrated in a few cities and metropolitan areas and it is this lack of spatial diffusion which creates the phenomenon of prosperous and depressed regions. However, since the growth process is cumulative and self-reinforcing, those cities which have gained a head start will be able to parlay this initial advantage into a situation in which they soon outstrip their rivals in terms of economic growth and in-migration. The laggards, on the other hand, will languish precariously between a state of slow growth and imminent decline. But clearly, this process cannot continue forever. Otherwise, most economic activities would agglomerate in a few major centers and new cities would find it almost impossible to enter the upper echelons of the urban hierarchy. The rise of the Sunbelt would thus be a theoretical impossibility.

Persky (1973:20), in an article which describes the historical evolution of the South in terms of its dependent favored colony status vis-à-vis the industrial Northeast, falls into just such a trap when he argues that economic divergence between the South and the North is "much better explained by a theory which assumes that a favored colony is in an excellent position to receive the spin-off of older industries from the metropolitan center, but not to generate or quickly partake in the dynamic phases of innovative cycles." While we do not quarrel with Persky's assessment of the pre-World War II relationship, it would appear that his attempts to project this situation into the present represents a stubborn adherence to an oversimplified version of the uneven development theory which no longer conforms to reality. The South may at one time have been locked into a dependent colonial status which ensured that the growth of the North would proceed in tempo with the increasing misery of the South. However, this relationship has since been terminated and, indeed, would appear to have been reversed as well. In other words, at present, Southern prosperity may be a significant factor contributing to Northern decline. Thus, it is incumbent upon adherents of the uneven development theory to refrain from a Procrustean application and instead recast the theory to include the economic development of the Sunbelt and the decline of the Northeast.

To restate the paradox, uneven development postulates that capital accumulation proceeds in a spatially concentrated manner. Once a certain critical mass is achieved, growth in the favored region proceeds in accordance with Myrdal's theory of cumulative causation. This in turn produces regional polarization or divergent development, i.e., the rich region prospers to the detriment of the poor area. However, at least within the U.S., it would appear that some mechanism has caused the regions to switch roles. The concept of divergent development still remains operative; only the role of leader and laggard has been reversed. The problem is to identify the mechanism which produces the switch.

One facet of the solution rests with the fact that previously dynamic industries eventually encounter a period of growth retardation and possible obsolescence (Burns, 1934). Their ability to provide additional thrust to the local economy is soon exhausted and this brings the period of rapid urban growth to a close. But while the growth rates in the initially dynamic industries are ebbing, the national economy is developing a new set of dynamic industries. The question is where will these new industries locate. If they locate in the old and now slowly growing regions, the local economy of these cities will be rejuvenated and launched upon a second round of cumulative growth. However, there is another alternative. They can forsake the previously developed area in favor of the backward less developed region. If the new dynamic industries pursue this path, then there will be a reversal of the role of leader and laggard within the cumulative growth process. But why would a new industry select a new region? Traditional economic theory would suggest that lower factor costs act as a powerful incentive. While there is certainly no reason to doubt that cost minimization considerations play an important role in the site selection process, we would suggest that the spatial evolution of the U.S. economy is governed by a more complex set of forces.

Specifically, the waves of urbanization which create new rapidly growing cities can be correlated with fairly distinct phases of capital accumulation. Within each epoch qualitatively different economic activities predominate. Thus, during the mercantile phase, national and urban prosperity was defined in terms of the success or failure of various commercial ventures and their ancillary, generally craft-based, manufacturing activities. Once this phase of capital accumulation had spent its developmental potential, a new epoch emerged to carry the national economy to new heights of prosperity. This second developmental era was based on the creation of the capital infrastructure needed to carry the industrial revolution to completion. During this second epoch, commercial activities did not disappear nor did they lose their importance in some older cities. They simply receded into the background, no longer serving as the foundation for the prosperity of the *newly* emerging urban centers. This cyclical pattern was repeated a third time when the industrial epoch gave way to a new constellation of dynamic activities founded upon the pillars of mass consumption, services, and the military industrial complex. This present phase is often identified euphemistically as the post-industrial society.

The concept of phases of capital accumulation only represents one aspect of the complex array of forces which impinge upon the patterns of spatial development. To complete this discussion, it is necessary to introduce the notion of developmental barriers. The concept of barriers, as employed in this paper, denotes a rather complex matrix of pressures and internally generated contradictions which serve to limit the development of a group of cities. The alternating influence of these barriers creates the shifting rhythms and patterns of uneven regional development.

Specifically, barriers operate in two directions; at one time they may be viewed as analogous to protective tariffs, while, at a later date, they serve as fetters blocking the city's adoption of a new set of dynamic activities. During the period before a specific epoch has spent its generative potential, the dominant cities erect various institutional and structural barriers which thwart the entry of other subordinate cities into the upper echelons of the urban hierarchy. By limiting the number of competing urban centers, the dominant cities inhibit the development of their less developed counterparts and thus make it more difficult for them to capture a significant portion of the epoch defining activities. This latter group of cities is relegated to the role of satellite or perhaps a favored colony. They are the hinterland outposts in an urban system whose development is controlled and regulated by the decisions emanating from the dominant cities. As such, their failure to grow and develop proceeds in tempo with the growth and prosperity in their dominant counterparts.

However, each epoch, if it has successfully fulfilled its developmental function, soon generates a set of contradictions which preclude further capital accumulation based upon the old and now obsolete practices. Thus, those cities which had previously crafted a set of barriers to insure their dominance in a particular sphere of economic activity become trapped by their own actions. Unless they can overcome their addiction to these old activities and transform their economies so that they are once again in harmony with the requirements generated by the succeeding wave of capital accumulation, their local economy will ossify and their position as growth leader will wane. The benefits flowing from their initial advantage during one developmental phase are thus transformed into barriers which inhibit the emergence of a new generative industrial base.

At this point, these barriers represent an opportunity for those cities which formerly had been relegated to a subordinate role. As one epoch wanes, a new and different set of possibilities for further capital accumulation appears on the horizon. Unencumbered by the now outmoded practices which characterized the previous era and which have immobilized the businessman wedded to the old methods and practices, these entrepreneurs craft a new set of dynamic cities based upon a modern approach to capital accumulation. In the process, however, they too erect a set of protective barriers only to be consumed by them at a later date.

In summation, within each epoch, the barriers erected by the dominant cities serve to produce divergent development. The dominant cities prosper by inhibiting the development of their potential rivals. However, the role of dominant and subordinate urban areas is not immutably fixed. Once the dominant cities encounter a set of internally generated barriers, their growth slows and opportunities arise for some of the subordinate cities to emerge from their colonial status and initiate a new wave of divergent development.

THE HISTORICAL PATTERNS OF DIVERGENT DEVELOPMENT

In this section, the preceding discussion of divergent development will be used as a guide for interpreting the history of the evolution of the U.S. system of cities. Four salient features of each developmental epoch will be discussed: the nature and impact of the epoch defining activities, the barriers which the leading cities erect to ensure that their rivals remain in a subordinate position, the internal contradictions which herald the end of a particular phase of capital accumulation, and the way in which these contradictions provide opportunities for a new set of cities to emerge based upon a new model of capital accumulation. However, one final word of caution is necessary before we proceed with the historical analysis of each epoch. These vignettes are not intended to provide exhaustive and definitive case histories either of specific cities or eras. Rather, they are meant to be suggestive. Hopefully, they will highlight some of the major variables and processes which a full-scale historical investigation should document in greater detail. Thus, they are presented as tentative beginnings, not as conclusive historical pronouncements.

URBANIZATION AND MERCANTILE ACCUMULATION

The history of U.S. urbanization is, in great measure, also the history of the development and evolution of U.S. capitalism. Capitalism, however, did not simply arise out of thin air nor did capitalist institutions immediately appear in their fully matured form. Instead, this mode of social and economic organization had to be meticulously constructed. Of paramount importance was the elimination of the self-sufficient frontier economy and its replacement by a commercial agricultural system. During the mercantile phase of U.S. economic development, those cities which spearheaded the assault on these barriers soon emerged as the leading metropolitan centers.

According to the economic historian N.S.B. Gras (1922:201), "mercantilism was the policy of the town writ large in the affairs of the state. . . . In the first phase of metropolitan economy, the chief concern of the metropolitan unit was not so much advance in industry as reorganization of the marketing system." Under mercantilism, each city is primarily concerned with carving out the most extensive hinterland possible in order to dominate the economic life of rival cities. The result is an economic landscape which can best be described as a "mosaic of resource regions" with each region dominated by a few major urban centers.

Two immediate obstacles—one attitudinal and the other physical—stood in the path of mercantile accumulation. With respect to the psychological barrier, it was noted that farmers could not be induced to produce a surplus above and beyond their own subsistence requirements unless markets were available. As early as 1810, Congressman David Porter of New York observed:

> The single circumstance of want of a market is already beginning to produce the most disastrous effect, not only on the industry, but on the morals of the inhabitants. Such is the fertility of their land that one-half their time spent in labor is sufficient to produce every article which their farms are capable of yielding, in sufficient quantities for their own consumption, and there is nothing to incite them to produce more. They are, therefore, naturally led to spend the other part of their time in idleness and dissipation. Their increase in numbers far from encourage them to become manufacturers for themselves, but put to a greater distance the time when . . . they submit to the labor and confinement of manufacturers. [Callender, 1902:23]

Although these psychological attitudes were anathema to mercantile interests, the self-sufficient mentality of the frontier farmer could not be replaced with a more commercial outlook as long as the prevailing transportation system retained its crude and primitive form. The significance of this transport barrier can be documented by examining the freight charges preceding the vast program of internal improvements.

In 1816, before the beginning of work on the Erie Canal, the cost of transporting one ton of goods from Europe to New York was $9.00. This same nine dollars would have paid for the transportation of one ton of freight 30 miles over the primitive turnpike system then in existence (Pred, 1973:114). European goods destined for Cincinnati were shipped inland via New Orleans and the Mississippi

River rather than overland through one of the East Coast ports. Neither high value manufactured goods nor low value, bulky farm produce could be sold at a profit under the prevailing system of freight rates (Taylor, 1964).

Not only were the internal freight costs prohibitive, but the time required for hauling goods overland was another serious obstacle blocking the establishment of intimate commercial ties. For instance, it required at least 50 days to ship commodities from Cincinnati to New York. A lag of this duration between shipment and arrival at the final destination meant that inland merchants and farmers could never be sure of the price and market situation. In a period of somewhat less than two months the market could become glutted and an entire crop sold for a loss. The risks were just too great to encourage regular commercial connections.

As a result, most of the hinterland commerce which did exist was controlled by the merchants of New Orleans. A symbiotic relationship developed between the West and the South, aided and abetted by those cities which were strategically situated at break-in-bulk points along the major waterways. If left unchecked, the development of western and southern cities would proceed at the expense of the East Coast port cities. However, the East Coast merchants were not content to sit by idly while this lucrative trade slipped from their grasp. Instead they sought to construct a comprehensive network of canals and railroads which, by reducing freight charges, would wrench the trade flows away from New Orleans.

With a concerted effort to construct canals and railroads, freight rates plummeted and trade patterns shifted dramatically. As early as 1851, one observer, J.D.B. DeBow, remarked that northern transportation systems had "rolled back the mighty tide of the Mississippi and its ten thousand tributary streams until their mouth, practically and commercially, is more at New York and Boston than at New Orleans" (Cochran and Miller, 1961:57).

But they did more than roll back the tide. By inhibiting southern growth, this new transport network fostered uneven development and allowed northern prosperity to occur simultaneously with deepening southern poverty. Fishlow (1964) reports that as early as 1839 the bulk of western produce was shipped directly to the major East Coast cities and was bypassing New Orleans. The decline of New Orleans as a transshipment center proceeded in pace with the completion of additional links in the internal improvement system.

"The proportion of flour flowing eastward or northward from Cincinnati increased from 3% in the early 1850s to 91% in 1860; similarly for pork, there was a shift from 7% to 42%. . . . Almost twice as much flour, eight times as much pork and bacon, twice as much lard, and three times as much of both corn and beef were exported from New Orleans in 1846-1849 than in 1858-1861" (Fishlow, 1964:355).

Trade flows along the Erie Canal corroborate Fishlow's assessment. In 1835, eastbound flour and wheat shipments amounted to only 268,000 barrels. By 1840, the shipments exceeded 1,000,000 barrels and, by 1860, they were 4,344,000 barrels. In 1838, flour receipts at Buffalo exceeded those at New Orleans.

Thus, while New Orleans did not suffer an absolute decline as a mercantile center, its growth had begun to wane and, with the completion of the canal system, the bulk of her exports consisted of cotton. The city was one of the first Southern centers to fall prey to an early round of divergent, uneven development as it lost the battle for control of the hinterland to the emerging mercantile centers of the Northeast. The canals had broken the barrier of north-south waterways and New Orleans, once the recipient of the benefits of these barriers, was now trapped by their elimination.

Despite southern attempts to retaliate with the construction of their own publicly supported internal improvements (Heath, 1950), they were fighting a losing battle. The trade routes had been permanently altered and the patterns of divergent development which would characterize southern economic history for the next 100 years had been firmly established. To many contemporary observers it appeared that southern wealth was the source of northern profits (Kettle, 1965). Southern complaints were loud and vociferous but to no avail. One southern writer, upset over the financial supremacy of the East Coast port cities, lamented:

> Last autumn the rich regions of Ohio, Indiana, and Illinois were flooded with banknotes of the eastern states, advanced by the New York houses on produce to be shipped by them by way of the canals in the spring. These moneyed facilities enable the packer, miller, and speculator to hold onto their produce until the opening of navigation in the spring and they are no longer obliged, as formerly, to hurry off their shipments during the winter by way of New Orleans in order to realize funds by drafts on their shipments. The banking facilities of the East are doing so much to draw trade from us as the canals and railways which eastern capital is constructing." [Albion, 1939:94]

Indeed, in the period immediately preceding the Civil War, mercantile interests had become so adept at controlling and financing this trade that they absorbed 42 cents out of every dollar of final sales (Marburg, 1960:555). Factors, acting as the agents of urban merchants, appeared in the major market centers to purchase the crops and arrange for their shipment. Inspectors, insurance agents, transport agents, and warehouse agents soon dominated the urban occupation structure and provided a major proportion of the total urban income. Most of this income was obtained at the expense of southern and western farmers and provided a major barrier to their development. Funds were drained from the agricultural regions in order to enrich the urban mercantile economies of the East. According to one Southern planter, "The steamer charges a dollar a bale. The sampler, weigher, drayman, piccory, warehouse, and pressman and brokers all have a snug percent. The factor has on average a dollar a bale for selling . . . and all that comes out of your pocket and mine" (Hasking, 1968:98).

In conclusion, at the outset of the mercantile era, the major developmental barrier was physical. Those cities which were most adept at overcoming this obstacle soon emerged as the nation's major mercantile centers. By directing trade flows through their borders, these cities inaugurated a process of cumulative and self-reinforcing growth fueled by the commercial profits which they were able to siphon from the agrarian regions. But with the passage of time, this matrix of mercantile activities lost its dynamic quality. The impact of this event profoundly influenced the development prospects in both the major mercantile cities and the subordinate urban outposts.

For the latter group of cities, the closing of the mercantile phase of capital accumulation meant that commercial functions were no longer capable of boosting them into the ranks of the major metropolitan centers. Unless they could surmount this obstacle by embarking on new methods of capital accumulation, they would be forever doomed to a subordinate status.

In the major mercantile centers, the waning efficacy of mercantile based growth had equally ominous consequences. During the mercantile era, these cities had risen to their present status by expanding their hinterland and eliminating the physical obstacles blocking the spatial extension of commercial relationships. But once each city's hinterland was firmly defined and the frontier penetrated and partitioned among the major centers, growth based upon further hinterland expansion was relatively difficult. Although it was unlikely that these cities would die, their potential for continued rapid growth based solely upon refinements of traditional commercial techniques and practices was clearly limited.

In both groups of cities, the termination of the mercantile phase of development was a direct outgrowth of mercantilism's previous success. By the end of this period, the physical isolation of the frontier had been eliminated, commercial intercourse between the frontier and the major mercantile cities along the East Coast and in Europe had been regularized, a national market had been created, and trade patterns had been successfully diverted from New Orleans. By dint of these accomplishments, the preconditions for a successful foray into industrial methods of capital accumulation were in place. The economy was poised for a new round of economic development based upon new ideas and processes. How the two groups of cities would react to these barriers/opportunities would determine the course of uneven development during the second phase of capital accumulation.

URBANIZATION AND THE INDUSTRIAL REVOLUTION

Although there has been some dispute concerning the date on which the U.S. economy embarked upon the industrial revolution, the influence of this shift on the urban system has been widely recognized for over a century. As far back as 1880, the census bureau emphasized that it is difficult to ascertain "in what proportion the growth of cities of the country as a body, has been due to commercial and in what proportion to industrial forces . . . but I conceive that no one will hesitate to assent to the proposition that the growth of cities of the U.S. since 1850 has been due in far greater measure to their development as manufacturing centers than to their increased business as centers for the distribution of commercial production" (U.S. Bureau of the Census, 1882:xxii).

The census bureau was not alone in its assessment. Many urban entrepreneurs and influential businessmen also recognized that the matrix of city building activities had shifted in favor of manufacturing. Kirkland (1961:163), for example, quotes one urban entrepreneur who declared, "The encouragement of the manufacturing industries is the most direct and most lucrative method of increasing the wealth of a community." The editor of the *Cleveland Leader* echoed these sentiments. "No thinking man with capital will stop here when we have only commerce to sustain us. A manufacturing town gives a man full scope for his ambitions" (Pred, 1966:18). The President of the Los Angeles Chamber of Commerce called for a program of "balanced prosperity" because, "It is our opinion that only by laying a high industrial foundation to our rapid growth can we hope to bring about a stabilized prosperity" (Fogelson, 1967:125).

The hue and cry for a manufacturing base to buttress the old mercantile functions was well founded. Not only had manufacturing activities grown dramatically during this period but the most rapid growth and the most dynamic industries were largely concentrated in the major urban centers. Any city which could capture a significant share of this new activity, therefore, could expect rapid growth and "stabilized prosperity." Generally, the most capital intensive operations, employing the greatest number of workers per establishment, and paying the highest annual wage were all located in the major urban centers. When total manufacturing data is disaggregated by industry, the same pattern is again observed (Watkins, 1977). In addition, between 1860 and 1890, the percent of total manufacturing employment concentrated in the largest 25 industrial counties rose from 30.2% to 42.4%.

However, despite the potential for a radical transformation of urban industrial structures wrought by the sudden prominence of manufacturing activities, many of the mercantile centers failed to capitalize upon this opportunity. Consequently, those cities which served as linchpins during the mercantile era saw their rates of population growth slow considerably. They were soon overshadowed by the rising younger cities of the industrial Midwest. This disparity in population growth rates is reflected in the different industrial structures present in the two groups of cities. For example, between 1880 and 1929, the major mercantile cities failed to substantially update their economic base. Mercantile activities were still their most prominent feature and, more importantly, the manufacturing industries dominant in 1929 were the same activities which held key positions in 1880 (U.S. Bureau of the Census, 1882, 1933). Very few old activities had been spun off and few new industries had been added. Moreover, most of the manufacturing industries present in the mercantile cities were among the slowest growing in the nation.

While the mercantile cities were limping along with such slow growing industries as printing and publishing, confectionary, clothing, bread and bakery products, furniture, and boots and shoes, the newly emerging cities had an industrial structure grounded in such industries as iron and steel, motor vehicle bodies and parts, tires, optical goods, machine tools, and so forth. The "newness" of these latter industries is apparent from their rapid growth. For example, employment in automobile bodies and parts rose from 1,810 in 1904 to 163,530 by 1923. From 1899 to 1923, automobile employment increased by *10,667.6%.* Other typical growth rates during this 24-year period are electrical machinery apparatus and supplies (459.1%); iron and steel, steel works and rolling mills (111.9%); iron and steel, bolts, nuts, washers, and rivets (108.6%); optical goods (220.3%); and rubber tires and inner tubes (406.9%) (U.S. Bureau of the Census, 1928). Thus, between 1880 and 1930, the industrial structure of the most rapidly growing cities was composed primarily of activities whose rate of employment growth far exceeded the national rate of manufacturing employment growth. These were the new dynamic activities generated by the industrial revolution and those cities which captured a significant share of these industries grew commensurably. The older mercantile cities, however, did not generate much employment in these new industries and, as a result, their growth was significantly below that of the industrial cities.

What caused these divergent patterns of development? What barriers prevented the mercantile cities from updating their economic base? Why were the newly emerging industrial cities capable of breaking through the barriers which had throttled the growth of the leading mercantile centers? To answer these questions we must consider each group of cities separately.

The failure of the dominant mercantile elites to recognize the efficacy of industrial practices has received considerable attention by economic historians. Mantoux (1961), for example, argues that inertia is a major concept which must be injected into any analysis of economic dynamics. Once a group of economic actors become enamored of one set of behavior patterns and mode of capital accumulation, they become loathe to foresake the proven effective methods of the past for new and untried approaches.

> The merchant manufacturers, accustomed as they were to the methods their fathers had used before them, found it hard to change. The outlay in equipment and building demanded by a factory frightened them. Why

> should they incur such charges when they could, or thought they could, earn just as much with less expense and fewer risks. The distance between them and the captains of industry was not great, but they never thought it worth their while to cover it. They soon had to bear the consequences of their timidity. [Mantoux, 1961:368]

If inertia was the dominant force preventing old cities from pursuing the revolutionary path of development, then it is fair to ask why this same phenomenon did not ossify the economic base in those cities which are now identified with the industrial revolution. In general, it would appear that two major factors stimulated the entrepreneurs in these cities to adopt behavior patterns which differed from those followed by their counterparts in the old mercantile cities.

First, those cities which were later to emerge as major industrial centers generally began their economic life as mercantile outposts which dominated relative small hinterlands. They served primarily as transshipment centers or break-in-bulk points for goods which either originated in the major mercantile centers or were ultimately destined for these cities. As such, they were subordinate cogs within the mercantile system. Their fate depended solely on the establishment of viable commercial links with one of the seaport cities and as a result, their prosperity and destiny was intimately related to decisions which were beyond their control. Their automony as mercantile centers, in other words, was severely constrained. However, with the emergence of a national market generated by the canal and rail networks, an opportunity arose for these entrepreneurs to liberate themselves from mercantile imposed barriers to development. If they could establish an independent economic base, then unfettered prosperity was a definite possibility.

In addition, many entrepreneurs who would have sought to emulate the eastern merchants realized that they had already lost the battle for the initial advantage. By trying to compete as mercantile centers, they would face the onerous task of catching up with those cities blessed with the advantage of an early start. However, manufacturing activities, especially in those lines which had not already been captured by the old cities during the mercantile era, presented an opportunity to ascend the urban hierarchy without competing directly with old cities. Thus, because of both their subordinate position within the mercantile system and the disadvantages associated with a late start, the entrepreneurs in these cities were less wedded to the old mode of capital accumulation.

The South, however, was not as fortunate as the industrial cities. Although the South, too, was in a subordinate position during the mercantile era, the emergence of industrial methods of capital accumulation did little to ameliorate its inferior position. In fact, as this second developmental phase progressed, the South became even more deeply mired in a state of relative and absolute economic deprivation. Not only were mercantile methods siphoning any investable surplus from this region, but now industrial practices were also employed to preclude self-sustained growth. The predicament of the South is vividly captured by the following quote from Henry Grady, the editor of the *Atlantic Constitution.*

> He gets up at the alarm of a Connecticut clock. Puts his Chicago suspenders on a pair of Detroit overalls. Washes his face with Cincinnati soap in a Philadelphia washpan. Works all day on a farm fenced with Pittsburgh wire and covered by an Ohio mortgage. Comes home at night and reads a Bible printed in Chicago and says a prayer written in Jerusalem. And when he dies . . . the South doesn't furnish a thing meant for his funeral but the corpse and the hole in the ground. [Persky, 1973:16]

Although Reconstruction certainly was a major political factor contributing to the South's dependent status, many other structural conditions within the economy also served as barriers blocking her development. Among the more significant were the post-Civil War agricultural system, absentee ownership of industrial facilities, and various freight rate arrangements.

That the South was laboring under all the disadvantages associated with a one-crop economy was recognized well before the Civil War. For example, in 1826 it was noted that, "There is not a finer grazing country in the world than South Carolina; and, were attention paid to the raising of cattle, sheep, goats, hogs, horses, mules, etc., this state might supply itself as well as all the West India Islands with these useful animals; but every other object gives place to cotton. Immense numbers of cattle, hogs, horses, and mules are driven annually from the western country into this state, and sold to advantage" (Callendar, 1902:127).

After the Civil War, the situation failed to improve. In South Carolina, for example, agricultural dependency on "foreign" products increased while, simultaneously, cotton production exploded. Between 1860 and 1880, the hay crop declined from 87,000 tons to

2,700 tons while cotton production increased from 353,000 bales to 522,000 (Woodward, 1951:178). In large part, the root of the problem can be traced to the fact that "at least half the planters after 1870 were either northern men or were supported by northern money." The link between northern money and southern agriculture was the lien system used to provide agricultural credit (Clark, 1946; Anderson, 1943). Under this system southern farmers in need of credit would pledge their unplanted crop as collateral to a merchant who in turn advanced the needed supplies. Once the lien was executed, the farmer was prohibited from purchasing on credit supplies from any other source and, in addition, could only sell the crop to his creditor. Since the merchant set the crop and supply price, the farmer was essentially forced to buy dear and sell cheap. After the interest charges were included, it was a rare occurrence for a farmer to "pay out" his debt in one season. Hence, he was forced to enter into a similar arrangement with the same creditor in the following year and, as a result, he "usually passed into a state of helpless peonage." With no recourse to other sources of credit, most of the farmers were forced to plant cotton—the one cash crop that the merchants would accept in exchange for credit—despite the fact that it was depleting the soil and glutting the market. Hence, these practices blocked the development of a more balanced agricultural system. Although northern merchants profited from this arrangement, Southerners sunk deeper into rural poverty.

While effective control of southern agriculture had passed to northern financial interests despite nominal southern ownership, Northern control of Southern industrial facilities was nearly absolute. According to Woodward (1951:292), "The Morgans, Mellons, and Rockefellers sent their agents to take charge of the region's railroads, mines, furnaces, and financial corporations, and eventually many of its distributive institutions. . . . The economy over which they presided was increasingly coming to be one of branch plants, branch banks, captive mines and chain stores." The effect was such that, in 1946, one northern journalist characterized Texas as "New York's most valuable foreign possession" (Danhoff, 1964:42).

The Birmingham steel industry presents a clear example of northern ownership throttling southern industrial development. In 1898, Birmingham was the world's third largest producer of pig iron and, in 1893, the South accounted for 22% of the nation's pig iron production. But between 1903 and 1913, pig iron production in the

South remained stationary while it expanded by 70% nationally. The failure of Southern production to expand is even more startling given the fact that most authorities concede that Birmingham mills had the lowest cost in the nation. Two related factors account for the moribund condition of the Southern steel industry. First, most southern steel capacity was acquired by the Morgan empire which, in 1907, amalgamated the Tennessee Iron and Coal Company with U.S. Steel. Second, the Pittsburgh Plus pricing scheme was instituted for the express purpose of concentrating steel production in the less efficient Northeastern areas.

Essentially, Pittsburgh Plus pricing required that all steel products be sold at a price determined by the mill cost at Pittsburgh plus delivery charges from Pittsburgh to the point of consumption. This mechanism effectively mollified the cost advantage of the Birmingham mills and encouraged all steel users to locate near Pittsburgh so that they could minimize freight charges. As a result of this pricing scheme, the geographic market for Birmingham steel products was severly limited. Stocking (1954:89) estimates that, as a result of this system,

> Birmingham mills supplied only 309 tons or 36.1% of the area's total requirements. They supplied no Texas buyers with drawn wire, although Texas bought more of it than any other state in Birmingham's natural market. Texans bought their wire largely from Pittsburgh, Sparrows' Point, Maryland, and Pueblo, Colorado. Louisiana customers bought 129 tons of plain drawn wire but they bought only five tons or less than 4% of their requirements, from Birmingham mills. Had Birmingham mills supplied all customer requirements in their natural market, and no customers elsewhere, they could have sold two and one half times as much were they in fact sold.

Many other southern manufacturing activities saw their markets constricted by the prevailing freight rate system which discriminated against southern producers (Potter, 1947; Heath, 1946). Although most evidence demonstrates that there were no cost differentials between Southern and Northern railroads, Southern producers were required by the ICC to pay higher freight charges for shipping their products a given distance than were Northern industrialists. As a result, a barrier was erected which protected Northern producers from incursion by Southern competitors.

THE RISE OF THE SUNBELT CITIES

In the previous two historical vignettes, the motivations of the major economic actors during each era were analyzed in order to assess the impact of their decisions on urban growth. We saw that the mercantile cities rose to metropolitan status by dint of the decisions made by a group of merchants attempting to create the infrastructure required for an exchange economy. Through their actions the last vestiges of a self-sufficient economy were destroyed and replaced by a national market economy. Actions in the sphere of exchange dominated the economy during this period and those cities controlling this development became the leading mercantile centers in the U.S. However, when these same economic actors proved unable to redirect the thrust of their activities, the local economies which they controlled stagnated relative to those of a new set of cities.

In the second wave of urbanization, the arena shifted to the sphere of production. Those entrepreneurs located in the subordinate tier of mercantile cities who recognized the potential inherent in industrial production guided their cities into the upper echelons of the urban system. In a sense, their failure to achieve an initial advantage in mercantile pursuits proved to be a blessing in disguise. Faced with a dependent secondary status and relatively freed from the inertial forces generated by the old and by now outmoded form of capital accumulation, these individuals had relatively little to lose by venturing into a new realm of capital accumulation. While the mercantile cities failed to develop a new set of dynamic manufacturing activities, these industrial revolution cities pressed forward by becoming the location for the new manufacturing city building activities.

This brings us to the third wave of urban development, euphemistically called the "rise of the Sunbelt." Traditional location and regional development theory suggests that the economic development of this region is a natural outcome of the free market process (Fuchs, 1962). Cheap land for space extensive manufacturing activities, low wage rates, low taxes, and a docile labor force all combine to make this region the most logical and cost-efficient location for both newly emerging sectors and old industries trying to maintain profit margins through a reduction in average costs. While there is no denying the veracity of these statements, we would argue that they are incomplete explanations which fail to address both the internal economic dynamics of these cities and the motivations impelling the dominant economic actors to behave as they do. In this section, a tentative and highly preliminary theory will be offered in the hope that further research will better illuminate the forces underlying this new wave of urban development activity.

In many superficial respects, this third era resembles the history of the previous two epochs. A set of barriers arose which precluded a continuation of the old industrial methods of capital accumulation. Out of these barriers emerged opportunities for those entrepreneurs and cities which could successfully catapult the economy onto a new growth trajectory. As the fate of various regions hung precariously in the balance, a struggle ensued over which region and set of cities would seize the opportunity. Would the Northeastern metropolitan areas be capable of overcoming the internal contradiction generated by the passing of the industrial era or would the mantle of growth shift decisively to a new set of cities? If the former situation prevailed, then the economy would witness a further intensification of the old pattern of uneven regional development as growth in the industrial cities produced increased poverty and underdevelopment in the South. But if the Sunbelt cities could capture a significant portion of the new epoch-defining activities, then the role of leader and laggard would be reversed and a new regional pattern of uneven development would emerge. All of these problems were not new. As the economy emerged from the stage of mercantile accumulation, the same constellation of issues, barriers, and opportunities were present, albeit in a slightly modified form. The economic adjustments required by these problems determined the course of uneven regional development during the industrial phase of capital accumulation.

But upon closer examination, the rise of the Sunbelt cities illuminates several factors which were absent in previous epochs and these additional nuances account for the contentious debate which embroiled the Sunbelt in a "second war between the states." Had the Sunbelt simply garnered a large proportion of the new epoch-defining activities, then the level of interregional dissension would have been moderate. However, in the present era, the rise of the Sunbelt represents a serious threat to the continued vitality and viability of the economic base of the Northeastern urban areas. Specifically, in previous epochs, the appearance of new entrants within the system of cities was not considered to be a precursor of death in the older cities. While local businessmen were distressed at the prospect of flagging growth, as long as the city could maintain its former economic base, absolute decline could be avoided. Uneven development was thus a relative proposition. One set of cities would lose their prominent position as the most rapidly growing urban areas and descend in rank through the urban hierarchy—although they would still be growing with respect to population and jobs—while another group assumed the role of growth leader. However, as the current indices of population migration and job loss indicate in the present wave of uneven urban development, absolute decline has replaced relative decline as the dominant motif. This in turn has raised the stakes for the growth laggards and heightened their sense of frustration.

Moreover, as if to add insult to injury, the rapid ascent of Sunbelt cities represents a dramatic break with past conditions. Not only have the northeastern cities lost their position as the preeminent national urban centers, but their role at the top has been usurped by a region which has overcome the deeply embedded subordination built up during two epochs of carefully crafted institutional and economic barriers. The Northeastern cities, once the recipient of the benefits of these barriers, found themselves trapped by their elimination. While they were saddled with an old, slowly growing industrial foundation, the Sunbelt cities, because they had previously been blocked from adopting these same activities, were in a more flexible position to shift with the changing needs of the economy. In a sense, they presented the economy with a *tabula rasa,* uncluttered with the outmoded infrastructure and habits characteristic of past eras. Moveover, the Sunbelt has seemingly erected a new set of barriers in the form of inequitable federal disbursements which, in the Northeast, are viewed as inhibiting any possible revitalization of the region. Thus, the tables have been turned and the previously dominant cities of the Northeast are witnessing an erosion, not only of their preeminent position in the urban system, but of their vital, life sustaining economic functions as well. While the Northeastern cities are losing population and jobs at an alarming rate, the Sunbelt is picking up the slack and leading the economy into new rounds of capital accumulation.

Before we briefly discuss some of the major factors producing this reversal of fortune in the South, it is important to dispel several interrelated myths which pass for explanations of the rise of the

South. Two in particular warrant special consideration. First, it is generally assumed that Southern industrial development has proceeded by luring low wage industries from their former location. While there is no doubt that the South attracts industries with the siren call of a docile, low wage, unorganized labor force, the concentration of low wage industries in this region has not been the major factor underlying its rapid economic ascent.

The dynamics of urban development in the South can be better understood if we first divide the manufacturing sector into two components. Those sectors in which employment declined or increased at a slower rate than the U.S. average for all manufacturing form the below average growth component. As of 1960, these sectors, with but one exception, offered an average hourly wage which was less than the U.S. average hourly manufacturing wage. The remaining manufacturing industries are assigned to the above average growth component and, as of 1960, they also comprised the above average wage segment. Thus, slow growing industries are generally associated with low wages and rapidly growing industries with high wages.

Once this division is made, several interesting points emerge. First, many of those sectors included in the below average growth category declined nationally although they grew by an average of 15% in the South. Thus, as a result of its lower pay scale, the South has captured a significant share of the low wage manufacturing industries and, in some activities, now accounts for as much as 60% to 70% of the national employment in these low wage portions of manufacturing. Since these same industries have been declining nationally while expanding regionally, the South has gained the reputation as a haven for the low wage components of manufacturing. But as of 1960, low wage industries accounted for only 40% of the South's manufacturing employment.

However, among fast growing, high wage industries, the South's performance was even more outstanding. Industries with an above average growth rate expanded by 92% nationally but by 180% in the South. Moreover, between 1940 and 1960, Southern manufacturing employment grew by approximately 1,560,000 jobs; almost 90% of this growth was centered in the above average growth, high wage group of industries.

Thus, it is misleading to attribute the emergence of the South simply to the relocation of low paying industries from the Northeast to the Sunbelt. This factor may account for a significant share of the decline recently experienced in the older metropolitan areas but it does not account for the great bulk of southern manufacturing growth. Clearly, this region is more than a low wage carbon copy of the industrial Northeast and its growth has not been limited solely to those low wage sectors which are looking for a cheaper labor site. Rather, because of its lower pay scales and less stringent labor laws, the South has emerged as a lower wage site for traditionally high wage activities.

A second, widely accepted myth argues that the rise of the Sunbelt is based primarily on pirating economic activity from the older industrial urban centers. In view of the widely advertised population and job losses in the Northeast coupled with rapid growth in the Sunbelt, this explanation would appear to accurately capture the dynamics underlying the interregional shift in the economic balance of power. However, despite its apparent veracity, this explanation fails to capture the essential process governing regional dynamics. Accepting this hypothesis would be tantamount to claiming that the industrial structure of Houston or Phoenix, for example, is nothing more than a relocated version of the economic base found in most Northeastern cities. Clearly, this is not the case. But if we fail to accept this scenario, we are left in an apparent quandary; how can we explain the dual phenomenon of growth in the Sunbelt coupled with decay in the Northeast without resorting to the overly simplistic assumption that the Sunbelt cities are simply replicating the industrial structure of their older Northeastern counterparts?

The solution rests with the way industrial activity diffuses throughout the various regions of the country. Specifically, an unindustrialized area can expand its economic base by any combination of three devices—physical relocation, branch plants, or the attraction of new leading edge activities. Depending on the mix, the region will either remain mired in its subordinate dependent status or it can break the barriers associated with the old mode of capital accumulation, initiate a new round of cumulative uneven development, and thereby reverse the role of regional leader and laggard. Southerners tried all of these devices at various times but soon realized that two of them—physical relocation and branch plants—were seriously deficient methods for achieving their goal of self-sustained development.

Physical relocation entails the actual movement of a production facility from one location to another. The former site is abandoned and a new location, blessed with a more favorable constellation of factor costs, is selected as a replacement. In the traditional parlance of urban economics, this phenomenon is known as industrial spin-off. However, several factors militate against lending too much credence to this process. Although physical relocation may produce adverse economic effects in the region or city which is losing economic activity, it is questionable whether the recipient region could reverse the process of uneven development solely by accepting those industries which have been cast-off by the more developed areas. As most studies of industrial spin-off indicate, industries which are spun off are generally mature, low wage activities (Thompson, 1965). As such, their growth is relatively anemic and unless the receiving region specializes in continuously absorbing large quantities of these activities, its growth will stagnate. Moreover, the low wages would fail to generate sufficient income to support all the ancillary nonbasic activities which are generally associated with a developed regional economy. Therefore, we would argue that if a region only receives the mature, low wage industries which are spun off from the more developed area, then it is an indication that the recipient has failed to overcome its colonial status. Although instances of physical relocation may generate significant attention, they are relatively inconsequential in explaining the rapid rise of the Sunbelt.

Branch plants can produce a shift in the regional alignment of industries in several ways—all of which obviate the need for physical relocation and hence are more difficult to detect. If a firm wishes to expand, then it can either build an additional plant in the old region or it can decide to construct the new facility in a new region. Similarly, if it already has several plants scattered in various regions then its decision as to which plant will receive additional capital infusions will also alter the preexisting regional balance of industrial activity. Finally, the firm can choose to abandon one of its branch plants while centralizing all activity in the remaining facilities. Irrespective of which situation actually prevails, the results will be the same; one region will grow while another either stagnates or declines.

However, in view of the traditional southern antipathy towards absentee ownership, it is unlikely that branch plants were a significant factor in the rapid ascent of the Sunbelt. While it is true that they are an excellent method for developing new industrial facilities in a region which is plagued by a shortage of indigenous capital, the dynamics of this process make it unlikely that the branch plant approach produced the phenomenal growth rates characteristic of the Sunbelt. In the first place, profit repatriation would ensure that any investable surplus accruing to these facilities would leave the region. This in turn would preclude the initiation of a substantial self-sustained growth process and as a result, the region receiving the branch plants would find that the pace of development is controlled by "foreign" investors. In short, "this way depends upon the initiative and decisions of companies outside the region. There is little that the region can do to speed up the process" (Hoover, 1949:27). Thus, if a region is dependent upon branch plants as the dominant source of its industrial development, this, too, is a signal that the region has failed to alter its subordinate status. Branch plants are a barrier, not an opportunity for reversing the role of leader and laggard within the context of a uneven regional development situation.

This leaves one other strategy by which an underdeveloped region can assume the role of growth leader. If new dynamic industries choose to locate predominately in the underdeveloped region while bypassing the old industrialized areas, then a new round of divergent uneven development will occur. According to Sale (1975) this process is primarily responsible for the rapid rise of the Sunbelt. In the postwar phase of capital accumulation, six new pillars of growth–agriculture, defense, advanced tehcnology, oil and natural gas, real estate and construction, and tourism and leisure–have emerged as the dominant industries. All have chosen a Sunbelt location. Northern cities, however, have failed to capitalize on the growth opportunities presented by these new industries and this has caused them to stagnate. Their inability to substantially update their industrial base as well as the adverse consequences resulting from this failure was recently underscored by John Dyson, the New York State Commerce Commissioner, who lamented, "What they don't know and don't see are the plants we don't get, the ones that are built in the South and the West. . . . As a result, our plants and equipment are getting more and more obsolete. They become less competitive and eventually they have to be closed, leaving people without jobs" (New York Times, February 2, 1976:1).

However, while we would not quarrel with Sale's description of the Sunbelt's economic profile, his work only addresses certain surface phenomena. It fails to go beneath these statistics to discover the root cause of the rise of the Sunbelt cities. For example, what prompted these six pillars to adopt a Sunbelt location? While Sale suggests that federal spending played a large role in their location decision and thus served as a powerful stimulus to urban development in this region, he never investigates the forces which impelled the federal government to assume such an active role. Certainly, the power of the Southern congressional delegation is a significant factor. In addition, the calculus of presidential electoral politics was another significant determinant of the federal largesse lavished upon this region. While the Democrats were wrestling with the attempt to keep the South in the Democratic column, the Republicans were fashioning a Southern Strategy to nullify the supposed Democratic monopoly (Billington, 1975). Thus, both parties had a strong incentive to direct federal spending in the Sunbelt. However, these conditions of political strategy and congressional alignment can be used to best advantage when coupled with a further analysis of the barriers and opportunities effecting continued growth based on the old industrial practices. Once we address this question, we will be in a better position to understand the economic raison d'etre of the Sunbelt and also to evaluate the current debate raging over the alleged bias in federal disbursement policies.

The immediate signal that the industrial wave had spent its generative potential was the trauma of the 1929 depression. However, although the outward manifestations of the problem first became painfully obvious in 1929, in retrospect, several economic historians place the first signs of trouble 10 years earlier. Both Kuznets (1961) and Creamer (1960) demonstrated that, beginning in 1919, the capital output ratio began a slow secular decline. They gave two reasons to explain this trend. First, they argued that up to 1920, large investments in fixed capital and infrastructure had been needed before production could commence. This raised the ratio of capital expenditures to dollar volume of output and was a major factor accounting for the buoyancy of the economy during the industrial epoch. Yet once these expenditures had been incurred, output could rise dramatically. As a result, the need for further increments of plant and equipment was reduced and the investment component of aggregate demand declined as a proportion of Gross National Product. Second, they argued that the rising burdens of depreciation expenses induced businessmen to economize on the use of capital. This was accomplished by developing new capital-efficient technologies which in turn produced the undesired side effect of a falling level of investment.

As the economy descended into the depths of depression, it was painfully obvious that the old business practices would not be adequate to rectify the present malady. Unless new modes of capital accumulation could be developed which would provide opportunities for the economy to break through the barrier of insufficient aggregate demand the prospects for further growth in the old industrial cities would remain bleak. Thus, according to both Kuznets and Creamer, the market for capital goods was the primary barrier blocking continued economic development. Until the investment component of aggregate demand could be raised to a more adequate level, the economy would continue to stagnate. However, as we will demonstrate the elimination of this barrier through the advent of Keynesian-inspired federal expenditure policies provided the opportunity for the emergence of a new phase of capital accumulation and its concomitant new wave of urbanization. The region which benefited most from these demand-stimulating programs was the Sunbelt, while the biggest losers were the old mercantile and industrial central cities of the Northeast.

In particular, federal policies to bolster aggregate demand coincided with the developmental needs of the Sunbelt in several distinct ways. First, if the Sunbelt was to pose a serious competitive threat to the economic dominance of the Northeast, its woefully inadequate infrastructure would have to be improved. Without this essential ingredient, the latent attractiveness of the Sunbelt could not be exploited. For example, irrespective of the prevailing low wage levels, absence of unionization, availability of energy, and politically hospitable environment, no firm will locate in a region unless the requisite social and physical overhead capital is already installed. Thus, if the Sunbelt desired to attract new dynamic activities, its first task would have to be the construction of a more profitable infrastructure.

At this point, the needs of the Sunbelt were in harmony with the demand stimulating policies of the federal government. If Northeastern central cities could no longer profitably absorb additional increments of urban infrastructure, then federal spending priorities could be shifted to new arenas. Duplicating these infrastructural facilities, in the Sunbelt and the suburbs, required massive public outlays for road construction and the expansion of ancillary support services such as electric, sewage, and water facilities. In addition, these expenditures created new and fertile areas for the private sector in the form of expanded opportunities to construct stores, shopping plazas, and office complexes. When coupled with the large multiplier effects generated by these projects, the lagging investment sector of the economy was given a powerful stimulus while the Sunbelt received the longed-for rejuvenation of its built environment.

Moreover, the design considerations which accompanied these projects provided an additional stimulus to aggregate demand. For example, the low density settlement patterns fostered a dependency on the automobile while all but eliminating the possibility for the installation of more efficient and hence less demand stimulating mass transportation systems. Also, low density, single family housing patterns required larger outlays for public services, construction materials, and operating costs (Council on Environmental Quality, 1974:9-12). All of the additional costs engendered by these settlement patterns helped to stimulate the economy and were at least implicitly sanctioned by various public expenditure and tax policies. To the extent that the uncluttered virgin territory in the Sunbelt was most amenable to these programs, the rise of the Sunbelt coincided with the requirements of the Keynesian phase of capital accumulation.

But even with the construction of an hospitably built environment, the Sunbelt was not prepared to assume the role of growth leader. In a sense, these preliminary steps were required and prefatory but they were not sufficient to boost the local economy. The missing ingredient was the adoption of a new dynamic industrial base. However, this too was provided as an offshoot of federal demand stimulating programs in the form of the military industrial complex. Electronics research and development facilities, calculators, semi-conductors, aeronautics, and scientific instruments–to name only a few of the most dynamic industries rising from the partnership between big business and big government–are all predominantly located in the Sunbelt. But without direct federal involvement in their inception and maturation, it is doubtful that their growth and economic significance would have been as dramatic.

Thus, in the third phase of capital accumulation, federal spending policies promoted both the construction of the needed infrastructure and the development of new dynamic industries. As a result, the developmental prospects in both regions were profoundly influenced by the inception of these policies designed to surmount the barrier of insufficient effective demand. The Sunbelt prospered from all phases of federal intervention and was able to assume the mantle of growth leadership. The Northeastern central cities, however, were not so fortunate and, thus, they lost their position as growth leaders within the U.S. system of cities.

However, the current controversy embroiling the Sunbelt in a "second war between the states" goes beyond a mere recitation of relative growth rates and the location proclivities of rapidly expanding industries. According to most accounts, the main point of contention concerns the alleged bias in federal disbursement policies which act as a barrier blocking the rejuvenation of the moribund Northeastern economies. Thus, at this point, it seems appropriate to briefly recapitulate and reassess the validity of this argument.

Generally, the debate focuses on the question of equity. Is the Northeast receiving its fair share of federal spending or has the Sunbelt been receiving more than it deserves? Depending on the author and the source of the statistics, different judgments emerge from the debate (Jusenius and Ledebur, 1976; Moynihan, 1977; National Journal, 1976).

But, in many respects, the issue of equity has guided the discussion along systemically irrelevant avenues. Questions of equity have never been prominent concerns of capitalism and, whenever resource allocation questions arise, efficiency criteria always take precedence over equity evaluations. Only under serendipitous circumstances do the two coincide. In view of these systemic characteristics, it only weakens the case for the Northeast to insist that equity criteria assume a more prominent position in the resource allocation calculus. For example, procurement policies, to the extent they are guided by competitive bidding procedures among private corporations which are free to locate their production facilities at any site they choose, will automatically favor the Sunbelt. Since these sites have the benefit of cheaper labor, lower taxes, and less expensive land, it is to be expected that Sunbelt cities will excel in procurement endeavors. Thus, within the capitalist system, as long as these decisions appear efficient, they will continue to be sanctioned.

If we dismiss the issue of equity, what other criteria can be used to assess the interregional impact of federal disbursement patterns? Many of the same reports which are couched in terms of equity also calculate each regions's balance of payments with the federal treasury. Although the precise figures vary from study to study, one general conclusion remains invariant. They all agree that the Northeastern states pay more to the federal treasury than they receive in the form of expenditures for contracts, wage and salary disbursements, and grants-in-aid. Conversely, the Sunbelt is favored with a balance of payments surplus. Since it is an economic axiom that prolonged payments deficits are deflationary, many of the proponents of restructured federal spending patterns use these deficits to buttress their case. However, even these arguments, although theoretically correct, are empirically tenuous. Until the precise regional patterns of subcontracting are ascertained, it will be impossible to determine the deflationary and inflationary impacts of interregional federal expenditure allocations. Thus, this argument also cannot be used to support the case for a policy reevaluation.

Yet there is one additional argument which, to this point, has not been pursued. Specifically, if we consider the marginal impact of federal disbursements and ignore both the regional balance of payments figures and per capita levels of federal disbursements, then a strong case can be made for a reconsideration of regional spending priorities. As we have tried to demonstrate, past federal policies have been instrumental in luring new dynamic industries to the Sunbelt. The impact of these policies is probably the single most significant factor contributing to the sudden ascent of these cities. At the same time, federal spending, irrespective of its absolute magnitude, has been relatively less successful in improving the industrial vitality of the Northeastern central cities. In other words, the marginal impact has been grossly uneven between regions.

The result of these marginal impact disparities has been absolute population and job losses in the Northeast coupled with rapid growth in the Sunbelt. However, people do not rush pell-mell across a continent without a reason. With the exception of retirees searching for a more desirable life-style and climate, people generally migrate to regions which promise the greatest job prospects. If jobs are available in the Northeast then surely the population exodus would moderate, if not cease. The issue, then, must be one of job creation and rejuvenation in the Northeastern central cities. Either federal spending can continue to promote the establishment of new dynamic industries in the Sunbelt or it can take steps to distribute these activities more equally. Pursuing the former policy may be tantamount to simply throwing away the old industrial cities in the hope that their Sunbelt counterparts can carry the burden by themselves. The latter policy provides an avenue to preserve these cities and perhaps end the historical patterns of uneven development whereby each region prospers by the misery of its rivals.

CONCLUSIONS

It is not the intention of this chapter to suggest that Northeastern cities be prevented from shrinking or that Sunbelt cities be inhibited from growing. No attempt has been made to render judgment on these issues. Rather, we have offered an approach to explain the economic shifts between the Northeast and the Sunbelt that is at once a theoretically specific analysis which places the two regions in a concrete economic context and an historically holistic explanation of the secular trends producing regional disparities.

In many respects, the convergence theorists, whose analyses we reject, harken back to the old theological debate concerning free will versus preordination. They argue that regional change and disparities are governed by uncontrollable forces of preordination. But instead of attributing all events to the will of an omniscient deity, they view the ineluctable operation of the Invisible Hand as the ultimate arbiter of all economic decisions. True, the Invisible Hand is posing problems for many of the older cities but these difficulties are part and parcel of a benign equilibriation process. As such, they suggest that no policies should be considered which would interfere with this adjustment since, after all the confusion is sorted out, all will be for the best in the best of all possible worlds.

In essence, this perspective views cities and regions as having a life and will of their own which supply the ultimate order for all human actions. From this perspective, human decisions do not positively influence the behavior of social phenomena. However poorly conceived, social science research would be of little consequence but for the fact that its ideas are used to shape the prevailing ideology and set limits on feasible policy alternatives. By deflecting attention from the true cause of the problems, this body of literature helps stabilize the prevailing status quo. Problems appear to be either the result of some inexorable forces operating independently of human control or else they are caused by defective people. In no instance are problems viewed as inherent qualities of a possibly defective social system. Consequently, any policy which advocates altering the societal institutions may be dismissed on the grounds that it is directing attention in a useless direction.

In our formulation of the regional development process, we hope that we have suggested an alternative which can dispel these illusory notions of social and regional dynamics. Throughout history, regional growth and decline has been governed by the decisions of individuals acting both in their capacity as private citizens and through various social institutions. The array of barriers and opportunities which have emerged to force a change in their prevailing economic practices have not had preordained impacts. Rather, during each stage, the fate of various cities rested on the perspicacity or lack of foresight of the dominant economic actors. Human decisions and not the Invisible Hand were the major controlling factor in the past just as they will continue to be preeminent in the future. Therefore, urban growth or decline must not be construed as inevitable. The economic fate of cities will always rest ultimately in the hands of the people.

REFERENCES

ALBION, R. (1939). The rise of the New York port. New York: Scribner.

ANDERSON, G.L. (1943). "The south and the problems of post civil war finance." Journal of Southern History, 9(2):181-195.

BILLINGTON, M.L. (1975). The political south in the twentieth century. New York: Scribner.

BORTS, G., and STEIN, J. (1964). Economic growth in a free market. New York: Columbia University Press.

BURNS, A.F. (1934). Production trends in the United States since 1870. New York: National Bureau of Economic Research.

CALLENDER, G. (1902). "The early transportation and banking enterprises of the states in relation to the growth of corporations." Quarterly Journal of Economics, 17(1):111-162.

CLARK, T.D. (1946). "The furnishing and supply system in southern agriculture since 1865." Journal of Southern History, 12(1):24-44.

COCHRAN, T., and MILLER, W. (1961). The age of enterprise. New York: Harper and Row.

Council on Environmental Quality (1974). Fifth annual report. Washington, D.C.

CREAMER, D. et al. (1960). Capital in manufacturing and mining. Princeton, N.J.: Princeton University Press.

DANHOFF, C. (1964). "Four decades of thought on the south's economic problems." Pp. 7-68 in M. Greenhut and W.T. Whitman (eds.), Essays in southern economic development. Chapel Hill: University of North Carolina Press.

FISHLOW, A. (1964). "Ante bellum interregional trade reconsidered." American Economic Review, 54(3):352-364.

FOGELSON, R. (1967). The fragmented metropolis: Los Angeles, 1850-1930. Cambridge, Mass.: Harvard University Press.

FUCHS, V. (1962). Changes in the location of manufacturing since 1929. New Haven, Conn.: Yale University Press.

GRAS, N.S.B. (1922). An introduction to economic history. New York: Harper and Brothers.

HASKING, R. (1968). "Planter and cotton factor in the old south." Pp. 92-112 in A. Chandler et al. (eds.), The changing economic order. New York: Harcourt, Brace and World.

HEATH, M. (1946). "The uniform class rate decision and its implication for southern economic development." Southern Economic Journal, 2(3):213-237.

——— (1950). "Public railroad construction and the development of private enterprise in the south before 1861." Journal of Economic History, 10(supplement):40-53.

HOOVER, C. (1949). Economy of the south. Report of the Joint Committee on the economic report on the impact of federal policies on the economy of the south. Washington, D.C.

JUSENIUS, C.L., and LEDEBUR, L. (1976). A myth in the making: The southern economic challenge and northern economic decline. Washington, D.C.: Economic Development Administration.

KETTLE, T. (1965). Southern wealth and northern profits. Montgomery: University of Alabama Press.

KIRKLAND, E. (1961). Industry comes of age. Chicago: Holt, Rinehart and Winston.

KUZNETS, S. (1961). Capital in the American economy. Princeton, N.J.: Princeton University Press.

KUZNETS, S. et al. (1960). Population redistribution and economic growth. Philadelphia: American Philosophical Society.

MANTOUX, P. (1961). The industrial revolution in the eighteenth century. New York: Harper and Row.

MARBURG, T. (1960). "Income originating in trade." Pp. 48-97 in J. Parker (ed.), Trends in the American economy in the nineteenth century. Princeton, N.J.: Princeton University Press.

MOYNIHAN, D. (1977). "The federal government and the economy of New York state." Congressional Record–Senate. June 27, pp. S10829-S10833.

MYRDAL, G. (1957). Economic theory and underdeveloped regions. New York: Harper and Row.

National Journal (1976). "Federal spending: The north's loss is the sunbelt's gain." June 26, pp. 878-891.

New York Times (1976). "States Commerce chief asks fiscal shift to right." February 2, pp. 1ff.

PERSKY, J. (1973). "The south: A colony at home." Southern Exposure, 1(2):14-22.

POTTER, D.M. (1947). "The historical development of eastern-southern frieght rate relationships." Law and Contemporary Problems, 12(3):416-448.

PRED, A. (1966). The spatial dynamics of U.S. urban industrial growth. Cambridge. Mass.: MIT Press.

——— (1973). Urban growth and the circulation of information. Cambridge, Mass.: Harvard University Press.

SALE, K. (1975). Power shift. New York: Random House.

STERNLIEB, G., and HUGHES, J.W. (1975). Post industrial America: Metropolitan decline and inter-regional job shifts. New Brunswick N.J.: Center for Policy Research.

STOCKING, G. (1954). Basing point pricing and regional development. Chapel Hill: University of North Carolina Press.

TAYLOR, G.R. (1964). The transportation revolution. New York: Holt, Rinehart and Winston.

THOMPSON, W.R. (1965). A preface to urban economics. Baltimore: Johns Hopkins University Press.

U.S. Bureau of the Census (1882). Report on the statistics of manufacturers, Volume 2. Washington, D.C.

--- (1928). The growth of manufactures, 1899-1923. Census Monographs VIII. Washington, D.C.

--- (1933). Manufactures: 1929. Volume 3. Reports by state. Washington, D.C.

WATKINS, A. (1977). Uneven development within the U.S. system of cities. Unpublished Ph.D. dissertation, New School for Social Research.

WILLIAMSON, J.G. (1965). "Regional inequality and the process of national development." Economic Development and Cultural Change, 13(4):3-84.

WOODWARD, C.V. (1951). The origins of the new south, 1877-1913. Baton Rouge: Louisiana State University Press.

2

Class Struggle and the Stages of American Urban Development

DAVID M. GORDON

□ WE HAVE BEEN ENTERING A NEW ERA in the history of American cities. Sunbelt cities are on the rise. Northeastern cities are waning like the old moon. The movements seem as inexorable as the tides. Why?

Many people seem to regard these changes as inevitable. For a variety of intuited but unelaborated reasons, many seem to assume that the Sunbelt era has come because Sunbelt cities are somehow "more efficient." It would be a mistake to resist the tides of change, in this view, because the modern calculus of efficiency has sanctioned them.

I disagree with this conventional view. I do not believe that the rise of the Sunbelt cities has been inevitable or that it is irresistibly necessary on narrow grounds of cost efficiency. Writing within the Marxian tradition, I argue instead that this recent shift in urban fortunes has flowed out of and reflects a long history of interaction between capitalist development and urban form in the United States. Like earlier transformations of urban structure, the recent moves to the Sunbelt have their roots in capitalist class relations and class struggles. In this essay I try to develop both the basis for and the implications of that argument.

AWAY FROM SPATIAL DETERMINISM?

The new urban form seems to be inexorably required by advanced industrialism. We may not like our lives in cities, but we seem to think that cities must continue to develop as they have if we are to maintain our present standards of living. How can we forget the need for coordination, for economies of scale, for urban agglomerations, for urban amenities!

This fatalism about our cities resembles an analogous fatalism about technology. That view is often called *technological determinism.* It suggests that our dominant technologies are the *only* kinds of machines which will permit our current standard of living. We may not like the alienated, specialized hierarchical jobs associated with those machines, but we have to accept them as requisites of our current affluence.

Analogously, our views of cities are suffused with a sense of (what I call) *spatial determinism.* This view suggests that there is only one way of organizing economic life across space, generating only one set of community relationships, which is consistent with advanced industrial standards of living. We may not like those urban relationships, but we have to accept them in order to enjoy what we have. Let them eat concrete!

Recent political struggles and social analyses have begun to challenge the technological determinist position. Research has suggested that modern machinery has not only permitted affluence but has also been conditioned by the particular characteristics of capitalist accumulation. (See Marglin, 1974; Braverman, 1974; and Gordon, 1976.) As Harry Braverman has put it (1974:230-231):

> These necessities are called "technical needs," "machine characteristics," "the requirements of efficiency," but by and large they are the exigencies of capital and not of technique. For the machine, they are only the expression of that side of its possibilities which capital tends to develop most energetically: the technical ability to separate control from execution . . . to subordinate the worker ever more decisively to the yoke of the machine.

This new view continually elicits puzzled responses from orthodox and Marxian social scientists alike. The dialogue typically reveals a common logical sequence:

Q: What generates the "werewolf hunger" for capital accumulation? A: Competition. Q: Won't competition force every capitalist to adopt the same production process? A: Yes, at the abstract level, because competition is the great equalizer. Q: Won't it also turn out that these forces and relations of production apply the "most efficient" technology and job structure? A: Yes, since those who did not employ the "most efficient" production process would lose in the competitive battle to those who did. Q: In that case, isn't it sufficient to speak of "efficient" technologies? What can we possibly add by speaking of technologies (or job structures) which "control" workers? A: Some technologies are better at controlling workers than others. Q: How can there be some technologies which differ from others? Isn't it true that there will be only one kind of technology adopted—the most "efficient" kind? How can there be some "more controlling" technology which is not "most efficient"? A: What you call most "efficient," I call "controlling." Efficiency is irrelevant by itself if the capitalist cannot produce surplus value.

Is this simply an impasse? Is the analysis of efficiency and control like comparing apples and oranges? Are the concepts incommensurable?

QUANTITATIVE AND QUALITATIVE EFFICIENCY

I have found it most useful to reformulate these questions within the following framework:

(1) In any class society, a mode of production can continue to dominate if and only if prevalent production processes reproduce the class relations defined by (the logic of) that mode of production. This requires a growth in the forces of production which is consistent with a particular pattern of class domination. It requires a set of social relations of production which reproduce ruling class power.

(2) In a specific class society, therefore, one cannot speak of the forces of production "in general." The forces of production are manifested in a historically specific production process. That production process either tends to reproduce ruling class dominance or tends to erode it. The same production process might have completely different effects in two different social formations.

(3) The "efficiency" of a production process, therefore, can be considered conceptually in two ways: efficiency has both a *quantitative* and a *qualitative* aspect.

In general, a production process is *quantitatively* (most) efficient if it effects the greatest possible useful physical output from a given set of physical inputs (or if it generates a given physical output with the fewest possible inputs). I can think of no theoretical reason why there would not be many (if not an infinite number of) possible production processes with equivalent quantitative efficiencies at any stage in the natural development of the means of production—in physical terms—in any given society.

In class societies, a production process is *qualitatively* efficient if it best reproduces the class relations of a mode of production. In more specific terms, a production process is qualitatively (most) efficient if it maximizes the ability of the ruling class to reproduce its domination of the social process of production and minimizes producers' resistance to ruling class domination of the production process. Given the opposition between the ruling class and direct producers, it would be surprising if production processes in a social formation *stably* dominated by a mode of production did not tend toward the most qualitatively efficient forms possible.[1]

With these three simple theoretical axioms, it seems possible to recast our archetypal Q & A exchange in more fruitful terms. Two frames of reference will prove helpful in the later discussion—historical and comparative static.

Historically, with the development of industrial capitalism, capitalists searched for any (among the many possible) quantitatively efficient production processes. Growing working-class resistance forced them to search for the most qualitatively efficient production processes. The search began on an exploratory basis. Those capitalists who discovered the most successful combinations gained comparative advantage over their competitors—not necessarily because their costs were minimized, in some prices-of-production sense, but because they were better able to discipline their workers, avoid strikes, and extract surplus product from their labor. These qualitatively efficient processes were copied, over time, and became more and more prevalent. Other quantitatively efficient processes fell away because they were less qualitatively efficient than those which triumphed. As capital accumulation continues, this process of exploration, discovery, and diffusion itself proceeds apace. As Braverman (1974) shows, its pace probably intensifies.[2]

There are two different ways to formulate this same kind of argument about capitalist production in *comparative static* terms—a weaker and stronger version.

More weakly, we can propose a simple formulation. At any moment in capitalist societies, given relative factor prices, there exists a set of quantitatively (most) efficient combinations of productive factors with more than one (and probably many) constituent elements. Within that possibility set, capitalism will tend to develop those production processes which are qualitatively most efficient. In more orthodox language, this formulation suggests that capitalist production processes maximize *qualitative* efficiency subject to the constraint that they are *quantitatively* efficient.

There is a stronger version toward which, I gather, some Marxists would lean. This version suggests that the criteria of quantitative and qualitative efficiency tend somewhat to conflict in capitalist societies. The requirements of labor control tend to push capitalists, according to this version, toward production processes which are less quantitatively efficient than other (technically) possible processes. This version suggests that, at least from time to time, the set of qualitatively (most) efficient production processes does not intersect with the set of quantitatively (most) efficient production processes. Capitalists are forced, in other words, to accept sacrifices in potential physical output, given prevalent scarcities and relative factor prices, in order to maintain worker discipline and reproduce their control over the means of production. In more orthodox language, in this case, capitalist production processes maximize quantitative efficiency subject to the constraint that they are qualitatively efficient.

This stronger version would help explain, for many, the apparent anomalies revealed in the literature on job design experiments. Virtually without exception, that literature suggests that worker productivity would increase if jobs provided more varied, less "degrading" tasks. And yet capitalists do not implement those kinds of job changes. One might hypothesize, in the theoretical language proposed here, that capitalists are unable to make quantitatively efficient changes in job structure because those changes would involve (relative) qualitative inefficiencies (see Gintis, 1976).

There are many problems with this kind of theoretical formulation. At best I think it serves to clarify some questions for further discussion and research. For the purposes of this essay, one simple implication seems most important: We can begin to apply the concept of *capitalist efficiency* in a very precise sense: production processes embody capitalist efficiency if they best reproduce capitalist control over the production process and minimize proletarian resistance to that control. As capitalism develops and as workers continually develop their organized capacity to resist capitalist exploitation, it seems logical that these imperatives of qualitative efficiency would become increasingly determining.

There is no reason why this analysis of qualitative efficiency should not be applied to urban form as well as to the labor process. We must begin to reconsider the spatial as well as the technological determinist view. The new view of technology has led us to the conclusion that capitalist machines develop at least partly in order to control us as workers. So may we also conclude, if we look closely enough, that capitalist spatial forms also develop at least partly to reproduce capitalist control, helping maintain the class relationships prevalent in capitalist societies.

This reconsideration clearly requires direct historical investigation. If we hypothesize, in Raymond Williams's (1973:302) words that "capitalism, as a mode of production, [has] created our kinds of city," then we must be able to trace the historical mechanisms through which these spatial consequences have gradually evolved. We must explicitly examine, in short, *the historical links between capitalism and urban development.* This essay explores those ties through a case study of American history.

ORTHODOX AND MARXIAN PERSPECTIVES

In turning toward that history, we quickly confront a conventional wisdom in the orthodox social sciences. Most urban histories treat the growth of cities as a gradual, evolutionary, and ineluctable process. The outcome seems destined. In any developing society, as Kingsley Davis (1965:9) writes with assurance, "urbanization is a finite process, a cycle through which nations go in their transition from agrarian to industrial society." Cities become continously larger, more complicated, more specialized, and more interdependent. Hans Blumenfeld (1965:42) describes the determinations of this historical process:

> The division of labor and increased productivity made concentration in cities possible, and the required cooperation of labor made it necessary, because the new system called for bringing together workers of many skills and diverse establishments that had to interchange goods and services. The process fed on itself, growth inducing further growth.

Because the United States has become the prototypical advanced industrial society, orthodox historians view American urban development as the consummate reflection of this universal process of urbanization.

In the Marxian view, that history must be seen in a different light. The Marxian analysis of the spatial division of labor suggests that no particular pattern of urban development is inevitably "destined," somehow deterministically cast in a general spatial mold. Spatial forms are conditioned, rather, by the particular mode of production dominating the society under study; they are shaped by *endogenous* political economic forces, not by *exogenous* mechanisms. Marxians also argue that urban history, like the history of other social institutions, does not advance incrementally, marching step by gradual step along some frictionless path. Urban history advances *discontinously,* instead of *continously,* periodically experiencing qualitative transformations of basic form and structure. During the capitalist epoch, in particular, the instability of the accumulation process itself is bound to lead to periodic institutional change. From this perspective, the current economic crisis and the attendant rise of Sunbelt cities are just another in a long series of these kinds of transformations.

This essay applies the Marxian perspective. According to that view, we have witnessed three main stages of capital accumulation in the advanced capitalist countries: the stages of *commercial* accumulation, *industrial* (or competitive) accumulation, and advanced *corporate* (or monopoly) accumulation. I argue that urban development in the United States has passed through *three corresponding stages*—each conditioned by the dynamics of capital accumulation which characterize that stage. I argue that the process of capital accumulation has been the most important factor structuring the growth of cities; city growth has not flowed from hidden exogenous forces but has been shaped instead by the logic of the underlying economic system. Finally, I argue that the transitions *between* stages of urban development have been predominantly influenced by problems of *class control in production,* problems erupting at the very center of the accumulation process.

This connected set of historical arguments can be separated formally into two main historical hypotheses about the relationships between capitalist development and urban form in the United States:

- First, that three principal urban forms have characterized urban development in America, each corresponding to a determinate stage of capital accumulation. In order to emphasize the *logic* of these connections, I give each urban form, defined abstractly, the name of its conditioning stage of capital accumulation, referring respectively to the *commercial city*, the *industrial city*, and the *corporate city*.
- Second, that the changes from one dominant urban form to another were forged in the crucible of capitalist production, determined by the struggles between owners and workers over social relations of production in the capitalist workplace, dominated by the criteria of qualitative efficiency.

I develop this argument in four additional sections. The first three trace the emergence of these three successive urban forms—the commerical city, the industrial city, and the corporate city. The final section applies that historical analysis to explore the current crisis of older American cities, arguing briefly that the relationship between capitalist development and urban form in the United States lies at the roots of the current rise of the Sunbelt cities.

COMMERCIAL ACCUMULATION AND THE COMMERCIAL CITY

During the final stages of the emergence of capitalism, merchant capitalists sought to increase their capital through *commercial accumulation*. Their profits depended on their capacity to "buy cheap and sell dear." They counted heavily upon political favors and franchises to strengthen their privileged intermediate positions in the marketplace. Because monopoly power in the market and political franchise through the state played such a critical role in determining the rate of commercial accumulation, tendencies toward *uneven development* among merchants, companies, cities, regions, or countries often emerged.

I would propose in general that cities served four kinds of political economic functions in this stage of commercial accumulation:

- A political capital (and colonial control centers) became the site(s) of the mercantilist government.
- The commercial metropolis housed the discounting, lending, accounting, and entrepreneurial functions supporting commercial exchange.
- Ports served as transport nodes—as centers for the collection and distribution of commodities being supplied from geographically diffuse points in the hinterlands and carried to dispersed markets.

- Since artisans producing luxury goods usually clustered in cities to gain access to their wealthy merchant and court-following customers, cities also served as craft manufacturing centers.

These few simple observations are sufficient to analyze the emergence and crystallization of the commercial city in the United States from the colonial era through the middle of the 19th century.

INTERURBAN DEVELOPMENT

Before Independence, American colonial cities served few economic functions. The first two general functions of commercial cities—those of the political capital and the commercial metropolis—were firmly lodged in London. Some seaports developed as transport nodes. Many artisans gathered in the colonial ports, producing luxuries for colonial merchants and administrators, but this craft manufacturing center function was obviously limited by the slow growth of the indigenous merchant class itself. The transport node function therefore dominated the distribution of people and economic activity between city and country.

Because colonial cities were serving such limited functions, the urban population did not grow very rapidly. Although total colonial population increased more than tenfold between 1690 and 1790, the cities' relative share of the colonial population actually declined from roughly 9% of colonial Americans in 1690 (Bridenbaugh, 1955:6) to only 5.1% in 1790.[3]

Given the singular importance of the transport node functions, there were strong political economic pressures which not only limited relative urban population growth but also limited the *number* of major cities. From the beginning, the British Crown strictly controlled town charters, fearing that British merchants would be unable to control commercial transport if too many port cities developed. In the North, these early political economic constraints limited the growth of commercial ports during the colonial period to just four places—Boston, New York, Philadelphia, and, somewhat later, Baltimore. The four ports accounted for 66.9% of total urban population in 1790, while only 50,000 lived in all other towns combined.

After Independence, the forces affecting urban development shifted. American merchants were able to gain control over a broader range of commercial functions, and American cities added the commercial metropolis functions to their earlier roles. As a result, the growth of American ports exploded. The urban share of total population *reversed* its decline during the colonial period; relative urban population rose from one-twentieth of the American population in 1790 to one-fifth in 1860.

With trade rising rapidly, urban merchants competed frenetically to gain control over both the commercial metropolis and transport node functions. During the first decades after Independence, the major ports were evenly matched in the battle for business. Between 1790 and 1810, the populations of Philadelphia, New York, Boston, and Baltimore each increased between 180% and 290%.

Soon enough, however, tendencies toward uneven development asserted themselves. New York merchants began to gain clear competitive advantage over their rivals. Between 1800 and 1860, the New York Port's share of total U.S. foreign trade climbed from only 9% to 62% (Vernon, 1963:31-32; Pred, 1966:147). By 1860, New York (including Brooklyn) contained over one million people, more than Philadelphia, Boston, and Baltimore combined.

INTRAURBAN FORM

The dynamics of commercial accumulation had much less effect on the *internal* structure of American cities. Because merchant capitalists were not yet intervening directly in production, commercial accumulation was not yet directly affecting the social relations of production. As a result, the port cities grew within the context of an earlier precapitalist set of social relations. In precapitalist cities, as Gideon Sjoberg (1965:8) writes, even business was "conducted in a leisurely manner, money not being the only desired end." As the commercial cities grew, they continued to reflect this earlier quality of random and leisurely life.

Within the central districts, most families owned their own property and acted as independent economic agents. Most establishments remained small, making it possible for nearly everyone to live and work in the same place. People of many different backgrounds and occupations were interspersed throughout the central city districts, with little obvious socioeconomic residential segregation. In the central port districts, the randomness and intensity of urban life produced jagged and unexpected physical patterns. Streets zigged and zagged every which way. Buildings were scattered at odd angles in unexpected combinations. It appears that a vibrant community life flourished throughout this central area. The cities featured "an informal neighborhood street life," in Sam Bass Warner's (1968:61, 21) words, threaded by the "unity of everyday life, from tavern, to street, to workplace, to housing."[4]

Only one group failed to share in this central port district life. Poor itinerants–beggars, casual seamen, propertyless laborers–all lived outside the cities, huddling in shanties and rooming houses. Too poor to establish themselves stably, moving frequently from town to town, they had little relationship to life in the urban center and little impact upon it.

By the beginning of the 19th century, each of the major ports increasingly reflected the characteristic structural logic of the commercial city. Each city was divided into two parts. One part coalesced around the wharf. Within that central district, many different occupational groups, filling the limited economic roles defined by the commercial city, lived and worked in intimate, intermingling, heterogeneous contiguity. The second part formed a band around the central port district. In it lived the transient homogeneous poor.

Once this basic form was established, rapid urban growth took place *within* it as long as the dynamics of commerical accumulation remained dominant. Between 1800 and 1850, trade expanded rapidly but industrialization had not yet dramatically invaded the urban scene. Some new occupational groups began to enter city life, but all of these groups continued to fit into the places and styles of the commercial city. The wharves continued to act as magnets, containing urban growth within the central port districts. People pushed toward the docks as much as possible. The central districts retained their heterogeneity and their precapitalist relationships; the ebb and flow of street life, however much more crowded it had become, seemed to continue. The poor also remained isolated on the outskirts.[5]

Cities did not become "modern" industrial cities, *until* capitalists began to turn toward a new mode of capital accumulation with a new set of characteristic relationships and struggles. Cities grew very rapidly, but they did not experience increasing physical separations among economic functions like work and residence. Nor did they exhibit increasing socioeconomic or ethnic residential segregation. Their poor people did not live in their central districts. And community life in the ports did not become demonstrably more impersonal and anomic. All that came much later.

THE CONTRADICTIONS OF THE COMMERCIAL CITY

This kind of emphasis may convey a misleading impression about the quality of life for early American urban residents; it may appear that I am romanticizing the "good old days" in the early American cities. Sharp conflicts did exist in the commercial city. Their basic contours developed directly from the central contradiction of the process of commercial accumulation itself.

Commercial accumulation tended to generate uneven development among buyers and sellers. This tended to effect, among other consequences, increasingly unequal distributions of wealth and income. Because different socioeconomic groups were living and working closely together in the commercial cities, these spreading inequalities became more and more physically evident—manifested in the luxury consumption of local merchants. It appears that this evidence of inequality generated popular protest against it. As inequalities reached their peaks both during the Revolutionary period and during the 1820s and 1830s, popular protests also seemed to intensify.

The recurrence of these kinds of protest helps preface the subsequent analysis of the industrial city. The dialectic of uneven development and popular protest reveals a fundamentally *spatial* aspect to the contradictions of the commercial path to capital accumulation. Capitalists sought a mechanism for increasing their profits on the surface of economic life—tracing profitable connections along the sphere of circulation. Their successes, manifested in increasing wealth, could not be hidden. Because the commercial city retained the precapitalist transparencies of immediate, intimate, and integrated social relationships, commercial capitalist profits could not be masked. The quest for such a disguise—the urgent need for which was so dramatically witnessed in the streets of the commercial city—assumed more and more importance as capitalists turned toward a new kind of capital accumulation.

INDUSTRIAL ACCUMULATION AND THE INDUSTRIAL CITY

Commercial cities were obviously superseded. City life had fundamentally changed by the end of the 19th century. Why?

Following our application of the Marxian perspective, we can draw our first clues from the pace and pattern of capital accumulation. In the United States, the years between 1850 and 1870 witnessed a transition from the stage of commercial accumulation to the stage of industrial accumulation. Capitalists turned more and more toward making profits through industrial production itself—through factory manufacture of the commodities which they exchanged on the market.

It hardly seems surprising that cities became the central locus for factory production. Cities provided easy access to markets, facilitating the scale of production necessary to support homogenized labor processes. Cities also provided easy access to pools of reserve workers, much less accessible to employers in the countryside.

But which cities would house those factories? And what would they look like?

THE TRANSITION TO INDUSTRIAL ACCUMULATION

The first major American factories—the textile mills of the 1830s and 1840s—were clustered along the rivers of New England in small cities like Lawrence, Lowell, Waltham, and Lynn. When coal replaced water as a source of energy between 1850 and 1870 and when railroads began to knit together the economic countryside, factories were freed from the river banks. Where would capitalists pursue their profits?

Huge cities eventually dominated as the loci for factories. Orthodox economic historians have argued that factories concentrated in large industrial cities for some combination of four main reasons: (1) they could be near large numbers of workers; (2) they could secure easy rail and water access to essential raw materials, particularly coal; (3) they could be near industrial suppliers of machines and other essential intermediate products, including "innovations"; and (4) they could be near consumer markets for their final goods. All of these factors are captured by the conventional term "agglomeration economies," and all revolve around criteria of quantitative efficiency.

But these hypotheses about agglomeration economies are not sufficient, by themselves, to explain continuing geographic concentration. At some point, increasing size and congestion can turn economies of scale into *dis*economies of scale.

Where is the threshold? When are industrial cities likely to approach the point of diminishing returns to economic agglomeration?

The relevance of these questions can be dramatized historically. Evidence suggests that the transition to industrial accumulation in the United States was witnessing a *diffusion* of urban factory location, not its centralization. In 1850, many cities of different sizes housed burgeoning factory production. As industrial capitalism took hold between 1850 and 1870, many of these cities enjoyed rapidly expanding industrial production regardless of their former size. Between 1860 and 1870, for instance, manufacturing employment in the three largest cities increased by only 53% while it increased in the cities ranked 21st through 50th in population by 79.5%. There was no significant correlation between rate of employment increase and initial population rank size (U.S. Bureau of the Census, 1850, 1860, 1870). The New Yorks, Chicagos, and Clevelands were growing rapidly, to be sure, but so were cities like Worcester, Jersey City, Indianapolis, and Dayton.

After 1870, as we know from hindsight, manufacturing began to concentrate in fewer and fewer large cities. By 1900, New York, Philadelphia, and Chicago each housed well over one million people. Manufacturing employment in those three largest cities grew by 245% between 1870 and 1900 while the number of industrial wage earners in the cities ranked 21st through 50th in population grew by only 158%. The 10 largest industrial areas increased their share of national value-added in manufacturing from under a quarter to almost two-fifths between 1860 and 1900 (U.S. Bureau of the Census, 1870, 1900; Pred, 1966:20).

How can we account for this rapid centralization of manufacturing employment without resorting to the kinds of ex post, relatively indeterminate explanations which the agglomeration hypotheses involve?

I hypothesize that a major reason for the concentration of manufacturing in the largest cities flowed from the dynamics of *labor control in production.*

Large cities appear to have become increasingly dominant as sites for capitalist factories because they provided a qualitatively efficient environment for capitalist factory production. Medium-sized cities did not fully satisfy this imperative. Larger cities satisfied it much better. And so, more and more capitalists built their factories in those large cities.

This hypothesis can best be elaborated through two separable questions: First, what differences between medium-sized cities and large cities account for their differential sustenance of capitalist production? And second, what explains those underlying differences between the two kinds of cities?

The first question is somewhat easier to answer. The problem of labor discipline plagued capitalists continually after they began to institute the factory form of production. Artisans resisted the degradation of work and wage laborers from preindustrial backgrounds struggled against the insecure wages and working conditions which factory homogenization continually imposed. Particularly as Civil War prosperity gave way to the stagnation and depression of the 1870s, workers fought to resist layoffs and wage cuts all across the industrial terrain.

In smaller and medium-sized industrial cities, employers had great difficulty suppressing and overcoming these moments of worker resistance. The power of the industrialist, in Gutman's (1963:11) words, "was not yet legitimized and 'taken for granted.'" Many middle-class residents, used to earlier, preindustrial relationships, resented the imposition of the relentless, uncompromising, impersonal disciplines of factory life. When workers struck, newspapers, politicians, and the middle classes often supported them. As Gutman (1963:48) elaborates, the nonindustrial classes "saw no necessary contradiction between private enterprise and gain on the one hand, and decent, humane social relations between workers and employers on the other."

In the largest cities, it appears, relationships among the several classes were significantly different. Workers were, in the transitional years at least, no less likely to strike than workers in smaller cities. But the various strata of the middle class were much more hostile to the workers than their peers in smaller cities. Newspapers, politicians, and the middle classes usually opposed workers on strike. Facing such hostility, workers found it more difficult to fight their employers. Because "there was almost no sympathy for the city workers from the middle and upper classes," workers were weakened and "employers in large cities had more freedom of choice than their counterparts in small towns." One of the many results, according to fragmentary evidence, was that "strikes and lockouts in large cities seldom lasted as long as similar disputes outside of these urban centers" (Gutman, 1963:41).

The implications of this hypothesis seem clear. Even if all other economic factors were equal—equalizing the factors of *quantitative efficiency* among cities—employers in large cities would be able to gain considerable advantage over their competitors in smaller industrial centers. If employers in large cities were better able to overcome worker resistance, they would suffer fewer losses during strikes, achieve greater discipline over their regular factory forces, and, in general, extract more surplus value from their workers. Even if employers were not particularly conscious of these relative advantages, at first, those located in larger cities would be able to grow more rapidly and profit more steadily than their classmates in other locations.

How can we account for these differences in class relationships and social environment between larger and smaller industrial cities?

Superficially, two explanations seem obvious. First, the greater physical segregation and impersonality of the larger cities seem to have isolated the working class and exposed it to community indifference or ostracism. Second, nonindustrial classes in smaller cities seem to have exhibited more militantly preindustrial values than their larger-city cousins.

These superficial differences are more difficult to explain. It appears that they need to be examined in different fashion for two separate groups of large industrial cities. In the Northeast, New York, Philadephia, Boston, and Baltimore—the four dominant commercial cities—became the four leading industrial centers in that region. Two characteristics of their precedent dynamics as the leading commercial cities seem to have played an important role in isolating their increasingly numerous industrial workers. First, they had already attained considerable physical scale as port cities and, on balance, had begun to suffer from the increasing impersonality which large scale tends to implant. Second, many of the professionals, politicians, newspapers, and merchants of those ports had begun to acquire—after years of support for commercial accumulation—a growing sympathy for the calculus of profitability.

Across the Appalachian Mountains, the most rapidly growing industrial cities displayed exactly the opposite traits. Those industrial cities, like Chicago, Detroit, and San Francisco, which changed character most quickly were those which *least* exhibited preindustrial community relations. Those midwestern cities, like Louisville, Cincinnati, Pittsburgh, and St. Louis, which had already developed commercial and preindustrial production activities before 1850, were the cities most likely to expose capitalist production to the friction of residual solidarities between workers and the middle classes. Indeed, if one looks at the 10 midwestern cities which were largest in both 1850 and 1900, one finds that those with the *fewest* wage earners in manufacturing in 1850 were precisely those whose industrial employment grew most rapidly between 1850 and 1900.

THE FORM OF THE INDUSTRIAL CITY

As the largest cities became increasingly dominant, the form of the industrial city crystallized. Its characteristics can be easily summarized.

First, huge factories were concentrated in downtown industrial districts, near rail and water outlets. Second, entirely new segregated working-class housing districts emerged. Located near the factories so that workers could walk to work, the housing was crammed together, typically clustered in isolation from the middle and upper classes. Although some ethnic segregation by block began to emerge, almost all working-class ethnic groups were contained within the same isolated areas (Hershberg et al., 1974). Third, the middle and upper classes began to escape from the center city as fast and as far as their finances permitted. Since the wealthy could afford to travel farther to and from work than the middle classes, residential socioeconomic segregation among those groups became more and more pronounced. The middle and upper classes were gradually arrayed in concentric rings moving along the transport spokes sticking out from the center.

Working from this schematic view, we can easily see that the industrial city represented a clear *reversal* of some of the most important tendencies reflected in the commercial city form. The central city was now occupied by dependent wage earners rather than independent property owners. Producers no longer worked and lived in the same place; there was now a separation between job and residential location. There was no longer residential heterogeneity; instead, the cities had quickly acquired a sharp residential segregation by economic class. In the commercial cities, the poor had lived outside the center while everyone else lived inside; now, suddenly, the poor and working classes lived inside while everyone else raced away from the center. In the commercial cities, central-city life involved nearly everyone in easy communality; in the industrial cities, only the working classes participated in the increasingly intense, impersonal, and assaulting street life, and they had little choice.

Was this new urban form destined by technical and spatial necessities? Apparently not. Other, much smaller, less segregated cities might well have facilitated equally rapid industrial growth. But capitalism requires workers' submission to their exploitation. Only in this kind of large industrial city, it appears, could workers be sufficiently isolated in order that their resistance might be rubbed smooth. As one foreign observer commented, upon surveying the central districts of Pittsburgh in 1884 (quoted in Glaab, 1965:236), "There are no classes here except the industrious classes; and no ranks in society save those which have been created by industry."

THE CONTRADICTIONS OF THE INDUSTRIAL CITY

Although the industrial cities grew rapidly, their growth did not proceed smoothly for long. As the end of the century approached, certain characteristic contradictions began to erupt. Accumulating friction began to threaten the speed of the industrial machine.

Most important, the latent explosiveness of the concentration of workers became more and more manifest. At first, through the 1870s, the impersonality and isolation of the factory and working-class districts had helped subdue the industrial proletariat. Gradually, through the 1880s, the dense concentrations of workers began to have the opposite effect. Individual strikes and struggles began to spread, infecting neighboring workers, Isolated moments of resistance took increasingly "political" forms. Strikes bred demonstrations not only at the plants but throughout the downtown districts.

The evidence for this relatively sudden intensification of labor unrest in the largest cities, spilling from one sector to another within the working class, seems reasonably persuasive. During the 1870s, most labor unrest took place in small towns, in the mines, and along the railroads. During the early 1880s, according to Florence Peterson's (1938:22) review of the data, "strikes were comparatively infrequent in the United States." From 1885, the magnitude, intensity, and form of strikes changed rapidly. (See Peterson, 1938:27ff.) In Table 1, the average annual index of workers involved in strikes is tabulated for five-year periods between 1881-1885 and 1901-1905. (The data series is interrupted in 1905.) The data reveal a

TABLE 1
WORKERS INVOLVED IN STRIKES, 1881-1905

Years	Index of Strikers, 5-Year Averages
1881-1885	56.8
1886-1890	118.6
1891-1895	125.8
1896-1900	124.0
1901-1905	187.4

NOTE: The index includes all workers involved in strikes for each year, 1927-1929 = 100, averaged over 5-year periods.
SOURCE: Peterson, 1938: 21, Table 1.

sharp increase in the numbers of workers engaged in strike activity during the 1880s and a steady quantitative increase after the first five years of the period covered.

Data on location of the strikes, although much more fragmentary, are also suggestive. Many strikes were taking place in the largest industrial cities. Two sources help suggest the general trends. First, strikes between 1881 and 1905 seem to have been concentrated in Illinois, Pennsylvania, and New York, the sites of the three major industrial cities; 59.8% of the workers involved in strikes over that period were located in those three states alone (Peterson, 1938). Second, increasing numbers of the major industrial disputes surveyed in the comprehensive labor histories of Foner (1955) and Commons (1918) were concentrated in the largest industrial cities.

As these contradictions began to erupt, it appeared likely that the form of the industrial city would have to change. Its original structure had been premised on its sustenance of capitalist control over production. The increasing centralization of the industrial proletariat which it promoted, however, was beginning to backfire. Labor control was threatening to dissolve. Something clearly had to give.

CORPORATE ACCUMULATION AND THE CORPORATE CITY

We now know that the industrial city was itself short lived. For about half a century, at least, our cities have been pushed in different directions. Corporate skycrapers have come to dominate the downtown districts of many cities. Factories have moved away from the central cities. Cities have become politically fragmented. And, finally, the Sunbelt city has emerged.

Once again, the argument of this paper suggests that we begin our analysis of these changes with an examination of the pace and pattern of capital accumulation. Around the turn of the century—between 1898 and 1920—the United States experienced a transition from the stage of industrial accumulation to advanced corporate accumulation. The accumulation process was being guided by the decisions of many fewer, much larger economic units. Those economic units—the giant corporations—now had sufficient size to permit a qualitatively new level of rationalization of production and distribution. Their size and scope led them increasingly to search for stability, predictability, and security. That search, I argue, played a central role in shaping the corporate city.

THE DECENTRALIZATION OF MANUFACTURING

Through the 1890s, as we saw, manufacturing had been concentrating in the largest central cities. Suddenly, around 1898 or 1899, manufacturing started moving out of the central city. In 12 of the 13 largest industrial districts in the country, a special census study showed that manufacturing employment began to increase more than twice as fast in the "rings" of the industrial districts as in the central cities. Between 1899 and 1909, central city manufacturing employment increased by 40.8% while ring employment rose by 97.7% (U.S. Bureau of the Census, 1909, 1910).

These numbers refer to a real and visible phenomenon noted by contemporary authors—in Graham Taylor's (1915:6) words, to "the sudden investment of large sums of capital in establishing suburban plants." Between 1899 and 1915, corporations began to establish factory districts beyond the city limits. New suburban manufacturing towns were being built in open space like movie sets. Gary, Indiana, constructed from 1905 to 1908, is the best-known example. Other new industrial satellite suburbs included Chicago Heights, Hammond, East Chicago, and Argo outside Chicago; Lackawanna outside of Buffalo; East St. Louis and Wellston across the river from St. Louis; Norwood and Oakley beyond the Cincinnati limits; and Chester and Norristown near Philadelphia.

Orthodox economic historians have conventionally explained the decentralization of manufacturing in the 20th century as the product of technological change—either the development of the truck and/or the diffusion of land-intensive automated processing machinery. But these conventional explanations cannot explain the sudden explosion of satellite suburbs at the turn of the century. The truck certainly had nothing to do with the development, since the truck was not an effective commercial substitute for freight transport until the late 1920s (McKenzie, 1933:93; Tunnard and Reed, 1955:238). There is no obvious evidence that there was a sudden rash of new inventions prompting a shift to land-intensive technologies; indeed, there is some evidence that the sudden decentralization took place *despite* shifts to *less* land-intensive technology in some industries.[6]

It appears that conventional economic historians have overlooked the major reason for the sudden dispersal of central city factories. Throughout the late 1880s and 1890s, as we saw above, labor conflict had begun to intensify in the downtown central city districts. Employers quickly perceived one obvious solution. Move!

In testimony presented before the U.S. Industrial Commission from 1900 to 1902, employer after employer explained the crystallizing calculus. Some examples (U.S. Industrial Commission, 1900-1902: VIII, 10, VII, 878, and VIII, 415, respectively):

> "Chicago today is the hotbed of trades unionism. . . . If it were not for the high investment [manufacturers] have in their machines and plants, many of them would leave Chicago at once, because of the labor trouble that exists here. . . . In fact, in Chicago, within the last two months we have lost some of the very largest corporations that operated here."
>
> Chairman of the New York State Board of Mediation and Arbitration:
>
> "Q: Do you find that isolated plants, away from the great centers of population, are more apt to have non-union shops than in a city? A: Yes. Q: Do you know of cases in the State where they do isolate plants to be free . . . from unionism? A: They have been located with that end in view."
>
> President of contracting firm in Chicago:
>
> "all these controversies and strikes that we have had here for some years have . . . prevented outsiders from coming in here and investing their capital. . . . It has discouraged capital at home. . . . It has drawn the manufacturers away from the city, because they are afraid their men will get into trouble and get into strikes. . . . The result is, all around Chicago for forty or fifty miles, the smaller towns are getting these manufacturing plants."

Graham Taylor, in his study of the satellite city movement, written in 1915, confirms that employers were particularly concerned about the contagiousness of central city labor unrest. The language of one of his examples (1915:23) is suggestive:

> In an eastern city which recently experienced the throes of a turbulent street-car strike, the superintendent of a large industrial establishment frankly said that every time the strikers paraded past his plant a veritable fever seemed to spread among the employees in all his work rooms. He thought that if the plants were moved out to the suburbs, the workingmen would not be so frequently inoculated with infection.

When factories did move to the industrial suburbs, Taylor (1915:101) notes, workers were automatically more isolated than they had been downtown: "Their contact with workers in other factories, with whom they might compare work conditions, is much less frequent." In general, Taylor concludes, the decentralization served its purpose and the unions were much less successful than they had been in the central city districts.

The great 20th century reversal of factory location, in short, began because corporations could no longer control their labor forces in the central cities. As with the transition to the industrial city, problems of labor control had decisive effects. U.S. Steel's creation of Gary metaphorically expressed the importance of this spatial effect. "The Steel Corporation's triumphs in the economics of production," Taylor (1915:227) concluded, "are only less impressive than its complete command over the army of workers it employs."

Two other major characteristics of the corporate city—the concentration of central business districts (CBDs) and political fragmentation of metropolitan areas—emerged after this first wave of factory decentralization and were permitted by it.

The first major expansion of downtown central business districts occurred in the 1920s. It appears that CBDs flowered in the 1920s because large corporations were not yet ready for them before then. Huge corporations had not consolidated their monopoly control over their industries until after World War I. Once they gained stable market control, they could begin to organize that control. They were now large enough to separate administrative functions from the production process itself, leaving plant managers to oversee the factories while corporate managers supervised the far-flung empire. Having already spurred the decentralization of many of their production plants, they could now afford to locate their administrative headquarters where it would be most "efficient." The uneven centralization of headquarters' locations quickly surpassed the concentration of industrial employment at its late 19th century peak. By 1929, according to McKenzie's (1933:164) figures, 56% of national corporations had located their headquarters in New York City or Chicago.

Timing is important for the explanation of political fragmentation as well. Conventional analysts explain 20th century political fragmentation as a consequence of residential decentralization. People began to prefer suburban autonomy, in this view, over central city domination.

The explanation does not work. Up to the end of the 19th century, central cities habitually annexed outlying residential districts as people moved beyond the traditional city boundaries. This process of annexation continued steadily until the end of the century. Then the expansion suddenly slowed. Chicago completed its last major annexation in 1889. New York City did not physically grow after the great incorporation of Brooklyn in 1898. Philadelphia and Boston had discontinued annexation even earlier. Of the 20 largest cities in 1900, 13 had enjoyed their last geographic expansion by 1910 (Jackson, 1972:443).

This rapid deceleration of central city annexation cannot be explained by some exogenous shift in people's preferences about suburban autonomy. People had been fleeing the central city since the 1860s. From the beginning, the refugees typically preferred autonomy and opposed annexation. Despite their opposition, extending suburban populations were simply reclaimed for the central city government by legislative *fiat*.

What changed at the end of the century? The most dramatic change, it appears, was that manufacturers themselves began to move out of the central cities. Obviously they wanted to avoid paying central city taxes. It was now in their interests to oppose further annexation. Given their influence over state legislatures, they easily satisfied their desires. Earlier residential opposition to annexation had not been strong enough to resist central city aggrandizement. Now, with manufacturers switching sides, the scales dramatically tilted. After industrialists joined the movement against central city extension, political fragmentation was the natural consequence.

THE FORM OF THE CORPORATE CITY

Once this transitional period had culminated in a stable pattern of urban reproduction, American cities had acquired a qualitatively new structure. It is reasonably easy to review the central political economic features of the corporate city.

If a city had reached maturity as an industrial city during the stage of industrial accumulation, its character changed rapidly during the corporate period although its physical structure remained embedded in concrete. Its downtown districts were transformed into downtown central business districts, dominated by skycrapers. (Because corporate headquarters were more unevenly distributed than 19th century industrial establishments, many industrial cities, like Baltimore, St. Louis, and Cincinnati, never captured many of these headquarters.) Surrounding the central business district were empty manufacturing areas, depressed from the desertion of large plants, barely surviving on the light and competitive industries left behind. Next to those districts were the old working-class districts, often transformed into "ghettos," locked into the cycle of central city manufacturing decline. Outside the central city there were suburban belts of industrial development, linked together by circumferential highways. Scattered around those industrial developments were fragmented working- and middle-class suburban communities. The wealthy lived farther out. Political fragmentation prevailed beyond the central city boundaries.

Many other, newer cities—particularly those in the South, Southwest, and West—reached maturity during the stage of corporate accumulation. They shared one thundering advantage over the older industrial cities: they had never acquired the fixed physical capital of an earlier era. They could be constructed from scratch to fit the needs of a new period of accumulation in which factory plant and equipment were themselves increasingly predicated upon a decentralized model. (Orthodox historians explain the decentralization of manufacuturing as a *result* of this new plant and equipment; I have argued that an eruption of class struggle initially prompted the decentralization and, by implication, that the new plant and equipment developed *as a result* of that dispersal in order to permit corporations taking advantage of the new locational facts.) There was consequently no identifiable downtown factory district; manufacturing was scattered throughout the city plane. There were no centralized working-class housing districts (for that was indeed what capitalists had learned to avoid); working-class housing was scattered all over the city around the factories. Automobiles and trucks provided the connecting links, threading together the separate pieces. The corporate city became, in Robert Fogelson's (1967) term, the Fragmented Metropolis. No centers anywhere. Diffuse economic activity everywhere.

These two models help underscore the significance of the *reversals* reflected in the corporate city form. Manufacturing had been clustering toward the center of the industrial city; now it was moving anywhere across the urban space. Working-class housing had been packed into dense central zones; now it was scattered around the metropolitan area and increasingly segmented. The middle and upper classes had been fleeing but were continually reabsorbed; now, in the older cities, they fled more successfully into separate suburbs. Before, the city had crammed around its center; now, the corporate city sprawled.

THE RISE OF THE SUNBELT CITIES

This argument helps place the rise of the Sunbelt cities, I think, in its proper historical perspective. It suggests that American cities have been dominated during the 20th century by either of two characteristic structures. Cities which reached maturity before the stage of corporate accumulation acquired what has become a relatively archaic physical structure. We can call these "old cities." Cities which reached maturity after the era of corporate accumulation had begun gradually developed a more "modern" physical shape. We call them "new cities."

Though shaped by the same underlying logic, old cities and new cities were bound to develop in different directions. New cities inevitably captured more and more manufacturing. Even the suburbs of the older central cities could not compete with the more perfectly suited physical environments of the newer cities, and industry has continually moved out of the older metropolitan areas–the old Industrial Cities–into the newer regions of the Sunbelt. Because corporate centralization continued, on the other hand, corporate headquarters continued to concentrate in a few CBDs. By 1974, although new cities had stolen huge chunks of manufacturing employment away from old cities, New York and Chicago still hosted nearly one-third of the 500 largest corporations' headquarters. In new cities, finally, annexation has never stopped. Industrialists had vested interests neither in the "original" central cities nor in the "expanding" suburban areas alone. Their economic interests were more or less equally distributed across the metropolitan plane. And so, given industrialist neutrality, opposition to central city annexation never developed in the new cities. The 20 most rapidly growing cities between 1960 and 1970 more than doubled their total land area between 1950 and 1970 (Jackson, 1972:441).

What explains the rise of the Sunbelt cities, in short, is very general and very simple. As newer cities, Sunbelt cities meet the tests of *qualitative efficiency* better than old cities. They have developed a form which lends itself to control of workers better than the older form. Mystified notions of the cost efficiency should not divert our attention from these original historical dynamics.

WHAT SHOULD WE DO ABOUT IT?

Over the longer term there should be nothing sanctified about the physical and institutional structure of *either* old cities *or* new cities. The forms of both were historically determined, in large part, by the needs of capital. Neither affords the basis for decent community life because both forms were historically conditioned by capitalists' efforts to isolate workers and then to divide them spatially. Rather than taking those urban structures for granted, we should begin instead to cast aside capitalist criteria altogether. The time has come to develop our own spatial forms.

NOTES

1. I am using qualitative efficiency in this context in a sense which is equivalent to Marglin's (1974) and Braverman's (1974) use of the term "control." I have chosen this formulation to emphasize the *connections* between considerations of "efficiency" and those of "control."

2. I have hoped that this formulation would make clear that discussion of qualitative efficiency does not necessarily involve reliance on "conspiracy theories." Class considerations can become determining without it having been true that all members of a class sat down together in a back room and conspired to achieve their common objectives.

3. Unless otherwise cited, all references in this paper to absolute and relative urban population and to individual city populations are based on U.S. Bureau of the Census (1939, 1976) and Bogue (1959).

4. This summary draws mainly on Bridenbaugh (1955), Warner (1968: Part I), Price (1974), and Pred (1966: Chapter 4).

5. For documentation, see the summaries in Warner (1968), Knights (1971), and Schnore and Knights (1969). As Schnore and Knights show, poor people continued to live on the outskirts of Boston at least through the 1840s. They also show that immigrants continually blended into the rest of the city without pronounced segregation through the 1850s.

6. In the steel industry, for instance, open-hearth furnaces were used more frequently in the Gary plants than in the central city plants but those furnaces required smaller plant units, rather than larger ones, when installed within the factories. See Clark (1929:68).

REFERENCES

BLUMENFELD, H. (1965). "The modern metropolis." Pp. 40-58 in Scientific American, Cities. New York: Knopf.
BOGUE, D. (1959). The population of the United States. Glencoe, Ill.: Free Press.
BRAVERMAN, H. (1974). Labor and monopoly capital. New York: Monthly Review Press.
BRIDENBAUGH, C. (1955). Cities in revolt. New York: Oxford.
CLARK, V. (1929). History of manufactures in the United States (vol. 1). Washington, D.C.: Carnegie Institution.
COMMONS, J. (1918). History of labor in the United States (vol. 2). New York: Macmillan.
DAVIS, K. (1965). "The urbanization of the human population." Pp. 3-24 in Scientific American, Cities. New York: Knopf.
FOGELSON, R. (1967). The fragmented metropolis: Los Angeles, 1850-1930. Cambridge, Mass.: Harvard University Press.
FONER, P. (1955). History of the labor movement in the United States (vol. 2). New York: International Publishers.
GINTIS, H. (1976). "The nature of the labor exchange and the theory of capitalist production." Review of Radical Political Economics, 8(2):36-56.
GLAAB, C. (1965). A documentary history of the American City. Glencoe, Ill.: Free Press.
GORDON, D. (1976). "Capitalist efficiency and socialist efficiency." Monthly Review, 28(3):19-39.
GUTMAN, H. (1963). "Workers' power in the gilded age." Pp. 38-68 in W. Morgan (ed.), The gilded age. Syracuse, N.Y.: Syracuse University Press.
HERSHBERG, T. et al. (1974). "The 'journey-to-work': An empirical investigation of work, residence, and transportation." Unpublished manuscript.
JACKSON, K.T. (1972). "Metropolitan government versus political autonomy." Pp. 442-462 in K.T. Jackson and S.K. Schultz (eds.), Cities in American history. New York: Knopf.
KNIGHTS, P.R. (1971). The plain people of Boston, 1830-1860. New York: Oxford.
McKENZIE, R.D. (1933). The metropolitan community. New York: McGraw-Hill.
MARGLIN, S.A. (1974). "What do bosses do?" Review of Radical Political Economics, 6(2):60-112.
PETERSON, F. (1938). Strikes in the United States, 1880-1936. Washington, D.C.: U.S. Government Printing Office.
PRED, A.R. (1966). The spatial dynamics of U.S. urban-industrial growth, 1800-1914. Cambridge, Mass.: Harvard University Press.
PRICE, J.M. (1974). "Economic function and the growth of American port towns in the eighteenth century." Perspectives in American History.
SCHNORE, L.F., and KNIGHTS, P.R. (1969). "Residence and social structure: Boston in the ante-bellum period." Pp. 247-257 in S. Thernstrom and R. Sennett (eds.), Nineteenth-century cities. New Haven, Conn.: Yale University Press.
SJOBERG, G. (1965). "The origin and evolution of cities." Pp. 25-39 in Scientific American, Cities. New York: Knopf.
TAYLOR, G.R. (1915). Satellite cities: A study of industrial suburbs. New York: Appleton.
TUNNARD, C., and REED, H.H. (1955). American skyline: The growth and form of our cities and towns. Boston: Houghton-Mifflin.
U.S. Bureau of the Census (1950). Seventh census. Washington, D.C.: U.S. Government Printing Office.

--- (1860). Eighth census. Washington, D.C.: U.S. Government Printing Office.

--- (1870). Compendium of the ninth census. Washington, D.C.: U.S. Government Printing Office.

--- (1900). Twelfth census, "Manufactures." Washington, D.C.: U.S. Government Printing Office.

--- (1909). "Industrial districts: 1905." Bulletin No. 101. Washington, D.C.: U.S. Government Printing Office.

--- (1910). Thirteenth census, Vol. X. Washington, D.C.: U.S. Government Printing Office.

--- (1939). Urban population in the U.S. from the first census to the fifteenth census. Washington, D.C.: U.S. Government Printing Office.

--- (1976). Historical statistics of the United States (rev. ed.). Washington, D.C.: U.S. Government Printing Office.

U.S. Industrial Commission (1900-1902). Reports. Washington, D.C.: U.S. Government Printing Office.

VERNON, R. (1963). Metropolis 1985. Garden City, N.Y.: Anchor.

WARNER, S.B. (1968). The private city. Philadelphia: University of Pennsylvania Press.

WILLIAMS, R. (1973). The country and the city. New York: Oxford.

3

Regional Change in the Fifth Kondratieff Upswing

WALT W. ROSTOW

□ RECENT REGIONAL TRENDS in the relative growth of population, real income, and industry, as well as North-South unemployment differentials, are to be understood in terms of three phenomena: the tendency of latecomers to modern economic growth to catch up with those who industrialized earlier; the differential impact on American regions of the first phase of what I have called the fifth Kondratieff upswing (explained below); and the overall inadequacy of national economic policy in dealing with the problems posed since the end of 1972. The failure to generate an effective national energy policy or to devise policies which would return the economy to sustained full employment has rendered our regional problems and tensions more severe than they had to be.

Economic historians have long been aware that, in the early stages of a country's development, regional differences in growth rates and real income are likely to increase. This happens because modern industrial technologies are picked up and applied, sector by sector, in areas endowed either with appropriate resources, location, or with particularly creative entrepreneurs. In the United States, for example, New England led the way with a modern textile industry in the 1820s; Pennsylvania with a modern iron industry in the 1840s and 1850s; Chicago with farm machinery; and, later, Detroit with automobiles. The northern regions of a number of countries industrialized before the south; in Britain, the United States, France, Germany, and Italy. Thus Stephen Potter's gamesman's ploy for breaking the flow of exposition at a cocktail party: "Ah, but it's different in the south." Brazil was an interesting exception: São Paulo and Rio led the way, while the Brazilian north, initially committed like the American South to a single crop (sugar), lagged. But with the passage of time and the availability of a diversified pool of unapplied technologies, the lagging regions begin to exploit their opportunities and to catch up. As one analyst of this problem (Williamson, 1965:44) concluded: "rising regional income disparities and increasing North-South dualism is typical of early development stages, while regional convergence and a disappearance of severe North-South problems is typical of the more mature stages of national growth and development."

What has been true within countries has also been true among countries. The familiar cliche that the poor get poorer and the rich get richer is simply not true. Both historical and contemporary evidence suggests the opposite. I have recently reviewed and analyzed that evidence (Rostow, forthcoming, a). For example, the average growth rate in income per capita of Britain, from the beginning of its industrialization in 1783 to 1967, was 1.3% per annum; for the United States, for the period 1843 to 1972, 1.8%. On the other hand, Japan, which started in 1885, and Russia, starting in 1890, both averaged 2.5%. Mexico, starting in 1940, has done even better: 3.4%.

The same broad result emerges if one looks at growth rates, in cross section, for the 1960s. The poorest countries (under $100 per capita in 1967 dollars) averaged only 1.7%. The rate rises steadily to a peak at about $1,000 per capita, where growth rates were 6.5%. The growth rate then declines for the richer countries, with the U.S. averaging in that decade only 3.2%.

The two basic reasons for this pattern are, I believe, these: first, the latecomers to industrialization have a large backlog of unapplied modern technologies to absorb, whereas the more advanced nations must depend on the flow of new technologies while carrying a heavier weight of old or obsolescent industrial plant; second, as countries (or regions) get richer, they allocate more of their income to services which, in general, do not incorporate technologies of high productivity to the same extent as manufactures.

The relative rise of the Sunbelt flows from the fact that it was late in moving into sustained and diversified industrialization and the modernization of its agriculture; but once the process took hold, the South moved ahead faster than the older industrial areas because it had a larger backlog of technologies to bring to bear. Its more rapid increase in income and accelerated urbanization amplified the process.

In the long sweep of the South's history, the turning point clearly came between 1930 and 1940, in the period of recovery from the Great Depression after 1929 (see Table 1).

Focus, for a moment, on the South Atlantic and East South Central states. Their relative income per capita declined between 1840 and 1860, as the North as a whole (as opposed to earlier New England) experienced its first rapid phase of industrialization; but, still, the southern regions stood at about two-thirds the national average in 1860, roughly the same level as the East North Central agricultural states. The Southern states lost ground seriously after the Civil War. This was a result not only of wartime destruction and the vicissitudes of Reconstruction but also because of a sharp decline in cotton prices and a slowing down in the expansion of the world's cotton consumption. There was some recovery from the mid-1890s to 1920, as the cotton price improved and some modest industrial

TABLE 1
PER CAPITA INCOME AS PERCENT OF U.S. TOTAL, BY REGIONS: 1840 to 1975

Year	USA	New England	Middle Atlantic	East North Central	West North Central	South Atlantic	East South Central	West South Central	Mountain	Pacific
1975	100	108	108	104	98	90	79	91	92	111
1970	100	108	113	105	95	86	74	85	90	110
1965	100	108	114	108	95	81	71	83	90	115
1960	100	109	116	107	93	77	67	83	95	118
1950	100	106	116	112	94	74	63	81	96	121
1940	100	121	124	112	84	69	55	70	92	138
1930	100	129	140	111	82	56	48	61	83	130
1920	100	124	134	108	87	59	52	72	100	135
1900	100	134	139	106	97	45	49	61	139	163
1880	100	141	141	102	90	45	51	70	168	204
1860	100	143	137	69	66	65	68	115	--	--
1840	100	132	136	67	75	70	73	144	--	--

SOURCE: 1840-1970, *Historical Statistics of the United States, Colonial Times to 1970*, Washington, D.C.: Department of Commerce, 1975, p. 242; 1975, *Survey of Current Business*, Vol. 56, August 1976, Table 2, p. 17.

development occurred in the South centered on the textile industry. In the 1920s the region again lost ground relatively, as agricultural prices sagged. In 1930 relative income per capita was only about half the national average. From the mid-1930s, however, four decades of sustained relative progress occurred. The East South Central states followed a similar path; although their relative income position in 1930 was a bit higher than in other parts of the South.

The rise of the Sunbelt in terms of relative income is, then, a phenomenon some 40 years old. It was accompanied by a large flow of black migrants to the North, set in motion by the push of cotton mechanization as well as the pull of jobs in the industrial cities. Between 1940 and 1970, for example, the white population of the South increased by 59%, the Negro population by only 21%. In the Northeast and North Central regions the Negro population more than tripled in these years, and increased more than tenfold in the West.

The process which brought about this striking movement of the South towards income equality with the rest of the country had these specific features:

- a remarkable decline in agriculture employment accompanied by the technical modernization of agriculture, including a shift to the West and mechanization of cotton production;
- a more than doubling of manufacturing employment, with a marked relative shift toward the production of durable as opposed to non-durable goods;
- a large shift of labor to construction as the region's population moved into the cities and suburbs (37% of the population in the South was urban in 1940, 65% in 1970);
- rapid expansion of public and private services, including education. In Texas, for example, the proportion of the population in 1940 in institutions of higher education was .6% as opposed to the national average of 1.2%: in 1970 the proportions were 3.1% and 3.4%. Much more than the gap in income was narrowed between the North and South in those thirty years.

In broad terms, then, the structural differences between the South and the rest of the nation were rapidly reduced over the period 1940-1970. As the converging employment data in Table 2 suggest, by 1975 it could no longer be said in the old way and to the same extent: "It's different in the south."

There were, of course, great differences among the Sunbelt states. Florida, for example, enjoyed a relatively high per capita income in

TABLE 2
PERCENT CHANGE IN NONAGRICULTURAL EMPLOYMENT AND LOCATION QUOTIENTS[a] FOR THE UNITED STATES, THE NORTHERN INDUSTRIAL TIER, AND THE SUNBELT-SOUTH: 1970-1975

	Percent Change in Employment: 1970-1975			Location Quotient			
				Northern Industrial Tier		Sunbelt-South	
Industry	U.S.	Northern Industrial Tier	Sunbelt-South	1970	1975	1970	1975
Total nonagricultural	8.6	1.3	16.7	––	––	––	––
Mining	19.6	13.6	21.5	.44	.40	2.0	1.90
Construction	– 2.2	–13.6	14.1	.84	.80	1.18	1.29
Manufacturing	– 5.2	–12.7	1.6	1.17	1.16	.96	.96
Transportation and public utilities	– 0.1	–5.7	9.0	.95	.98	1.02	1.03
Wholesale and retail trade	12.7	6.4	21.6	.96	.97	1.00	1.00
Finance, insurance and real estate	14.5	6.7	29.1	1.04	1.04	.90	.95
Services	20.4	14.6	30.8	.99	1.01	.88	.89
Government	17.3	11 8	23.5	.87	.89	1.06	1.04

a. A location quotient is the ratio of the proportion of a region's total employment which is located in a particular industry to the proportion of the nation's total employment which is located in the same industry.
SOURCE: U.S. Department of Labor, Bureau of Labor Statistics, *Employment and Earnings* Vol. 19 (May 1973), pp. 128-137; and Vol. 22 (May 1976), pp. 49 and 126-135. Compiled by C.L. Jusenius and L.C. Ledebur, "A Myth in the Making: The Southern Economic Challenge and Northern Economic Decline," prepared for U.S. Department of Commerce, Washington, D.C., November 1976, p. 34.

1940 (86% of the national average) and improved its position only modestly (1970, 94%), lacking either a substantial manufacturing structure or energy resources. It has remained disproportionately dependent on tourism. Primary metals are a major source of income only in West Virginia, Alabama, and Kentucky. Energy production is a major export of Kentucky, Oklahoma, Louisiana, Texas, and West Virginia. Other southern states face energy problems similar to those in the North. The proportionate role of agriculture, services, and industry as sources of employment varies greatly among the Sunbelt states. For example, 37% of employment in North Carolina is industrial, 13% in Florida. Mississippi, Arkansas, South Carolina, and Alabama remain still relatively poor states as compared, even, to the rest of the South and Southwest.

As of 1970, the uneven movement of the South toward rough homogeneity in income and structure with the rest of the country was a phenomenon little studied in the North; and, when noted, it was a source of gratification rather than anxiety. In particular, the success of the South in adjusting to the Civil Rights Acts of 1964 and 1965—ending Jim Crow and permitting a rapid increase in voting among southern blacks—was widely, if not universally, regarded as a victory for the nation as a whole in dealing with its oldest and most difficult problem.

Then came the fifth Kondratieff upswing.

Over the past several years I have been mildly bewildering various audiences and readers by suggesting that we are in the early phase of a fifth Kondratieff upswing. Who was Kondratieff and what is a Kondratieff cycle?

N.D. Kondratieff was a Russian economist. Writing in the 1920s, he suggested that capitalist economies were subject to long cycles, some 40-50 years in length. His views were published in the United States in summary in the mid-1930s. They generated considerable professional discussion and debate, but dropped from view in the great boom after World War II. Most contemporary economists vaguely remember having run across his name and ideas in graduate school but have forgotten precisely what it was he said.

Looking back from the mid-1920s, Kondratieff saw two and one-half cycles in various statistical series covering prices, wages, interest rates, and other data expressed in monetary terms. He sought but failed to find concurrent long cycles in production indexes. Kondratieff did not develop a theory of long cycles, but he asserted that a coherent explanation must exist. Since he wrote, the cycles he described have continued down to the present with a trough in the mid-1930s, a peak in 1951, a trough again in 1972.

My own explanation for the phenomena Kondratieff identified centers on shifts in the prices of food and raw materials relative to the prices of manufactures (Rostow, 1975, forthcoming, b).

The period since the end of 1972 is the fifth time in the past 200 years that a rise in the relative prices of basic commodities has occurred; and on each of the other four occasions it has been accompanied by manifestations similar to those we have experienced over the past five years: an accelerated general inflation, an extremely high range of interest rates, pressure on the real wages of industrial labor, pressure on those with relatively fixed incomes, and shifts of income favorable to producers of food as well as energy. The other four occasions occurred in the 1790s, the early 1850s, the second half of the 1890s, and the late 1930s. On each occasion, food and raw material prices then fluctuated in a relatively high range for about 25 years. Approximately another 25 years followed in which the trends reversed; that is, the prices of basic commodities were relatively cheap, as they were from 1951 to 1972. Each of these periods was, in an important sense, unique, and the trends did not unfold smoothly; but the fact is that the world economy for almost two centuries has been subject to a rough and irregular pattern of long cycles in which periods of about 20 to 25 years of high relative prices for food and raw materials gave way to approximately equal phases of relatively cheap food and raw materials.

I am not wedded to the notion that these cycles will continue in the future. But I would guess that the inexorable pressure of excessive population increase in the developing world, the tendency of the poor to spend increases in income disproportionately on food, the rising demand for grain-expensive proteins among the rich, the raw material requirements of a world economy where industrialization is spreading in the southern continents, and the high marginal cost of expanding the non-OPEC energy supply will persist for some time. Given these powerful and sustained forces operating on food, energy, raw material prices, and the costs we shall have to incur to achieve and maintain clean air and water, I believe we are in for a long period when the prices of these basic inputs to the economy will remain relatively high. Indeed, I would guess that we shall only have a fifth Kondratieff downswing if and when we create a new, cheap (hopefully infinite and nonpolluting) energy system. As we all know, energy is a critical factor not only in its own right but also because of its role in agriculture and in the extraction of raw materials and, potentially, in rendering economical the conversion of salt water into fresh water.

Down to 1914 the classic response to a Kondratieff upswing was to open new agricultural and raw material producing areas: the American West, Canada, Australia, Argentina, and the Ukraine. The great movements of international capital during this era were, in substantial part, induced by the price system, combined with new technologies of transport and production, to bring new supplies into the market and to restore balance in the world economy. In the fourth Kondratieff upswing (1933-1951), the diffusion of new agricultural technologies, rather than the opening of new physical frontiers, reestablished a tolerable balance in food production without much conscious government intervention; although the exploitation of Middle East oil after 1945 ranks, in the field of energy, with the opening of the American West in agriculture a century earlier. But in the 1970s and beyond we confront the fifth Kondratieff upswing period in a setting quite different from that of the past. I wish we could, but we cannot realistically rely to the same extent on the automatic workings of the price system and private capital markets to restore and maintain balance. All over the world, in one way or another, policy toward resources is in the hands of governments or is strongly influenced by governments. At every stage in the effort to restore balance, therefore, public policy will be involved. We shall have to think and consciously act our way through the fifth Kondratieff upswing.

It was the coming of the fifth Kondratieff upswing which suddenly converted a relatively benign pattern of regional development into something of a national problem. This was the case because the relative rise in food and energy prices accelerated the development of a good many Sunbelt states which exported energy and agricultural resources to the rest of the country; while the relative price shift decelerated the already slower rate of expansion in the Northeast and North Central industrial states.[1] The population shift to the Sunbelt picked up momentum; although about two-thirds of the recent population increase in that region is due to somewhat higher birth rates than in the North. More than half the nation's population increase between 1970 and 1975 was in the South. And the flow of blacks from south to north reversed.

In the sharp recession of 1974-1975, unemployment averaged in the latter year over 9% in New England, the Middle Atlantic, and East North Central states; 7.9% in the South Atlantic states, with Florida, dependent on tourism, as high as 10.7%; 6.9% in the South Central states. Texas experienced in 1975 only 5.6% unemployment. The relatively higher levels of unemployment cut tax revenues and increased requirements for compensating social services at precisely the time the tax base was also being weakened by the initial impact of the fifth Kondratieff upswing on real income and the accelerated flow of people to the South.[2]

Thus, the shift in relative prices since the end of 1972 did not create the problems of the Northeast and North Central states; but it converted a slow-moving erosion of their relative position into something of a northern crisis.

The role of the relative price shift in bringing about these changes is underlined by looking beyond the Sunbelt. Between 1970 and

1975 both the agricultural states of the West North Central area and the coal rich Mountain states also enjoyed a rise in relative prices, a favorable relative shift in income, and lower than average unemployment. The Mountain states, in fact, experienced the highest rate of population increase (16.3%) of any of the nation's regions over those 5 years.

The process at work since the end of 1972 on the pattern and pace of American regional development is quite familiar to students of international capital movements and migration in the past, although it has not yet found its way into the conventional literature on regional economics. Take, for example, the story of Canada from 1896 to 1913, a case I examined along with others (1975, forthcoming, b). A relative rise in the wheat price began from a trough in the mid-1890s. This produced a prompt improvement in Canada's terms of trade and rendered profitable the opening of new wheat lands. After the end of the Boer War, immense flows of capital from London supplemented Canada's capacity to import, already enlarged by favorable terms of trade and increased wheat exports. But the boom far transcended the expansion of wheat production. It was accompanied by accelerated immigration and Canada's first period of truly modern industrialization. On the other hand, Britain was burdened by unfavorable terms of trade and stagnant or falling industrial real wages, despite the prosperity of its export industries. The attractiveness of capital exports to Canada (and elsewhere) raised interest rates and cut into British domestic investment.

The impact of the relative price shift of the fifth Kondratieff upswing on the American regions is not precisely analogous, but it has involved a direct income shift associated with a change in the terms of trade; large movements of capital and people; a general economic acceleration in the areas favored by the terms of trade shift, a general retardation in those disfavored.

We turn now to the pattern of development since 1940 in the Northeast and East North Central states, and to the reasons for their peculiar vulnerability to the coming of the fifth Kondratieff upswing.

There is now a substantial consensus on the characteristics which have marked the evolution of these regions over the past three decades. Here, for example, is one statement of the situation of the East North Central region (Widner, 1976:i-ii).

"It is a region in trouble:

- National economic shifts are causing the region to lose more jobs than it is gaining;

- A substantial outflow of talent is now occurring as a result of outmigration toward the South and West;
- This outflow is leaving behind a large population of poor who migrated into the industrial cities in earlier decades;
- Many cities are dominated by one industry and have highly specialized, vulnerable economies;
- These conditions have led to substantial financial trouble as cities try to meet the rising costs of public services out of dwindling tax bases;
- These problems are compounded by fragmented government and taxing capacities in many of the metropolitan areas;
- These fiscal difficulties are reinforced by a highly unfavorable "balance of payments"—poorer than any other section of the country—with respect to federal taxes paid and federal funds received;
- It is at a disadvantage in competing with other sections of the country in terms of recreational and climatic amenities; and
- It is a region with a high percentage of obsolescence in its manufacturing plants, housing, and public facilities."

This array would constitute a conventional statement of the problems confronted in the Northeast—chronically in the 1950s and 1960s, acutely in the past five years. The Midwest has, however, an additional concern: the peculiarly important role of the automobile industry which ripples back from the assembly plants in Detroit to steel, rubber, and the other industries which furnish components. A 25% reduction in the demand for automobiles (or reduction in their size) might cause a loss of 400,000 jobs and increase unemployment in the region by 3.2% (Widner, 1976). The Northeast is somewhat more diversified. There is no single equivalent of this massive but now vulnerable industrial complex, although a good many cities throughout the Northeast are dependent on particular plants for a significant proportion of their employment.

The now-familiar relative shift of industry from the northern region is set out in Table 3, with its marked acceleration in the first half of the 1970s.

There is another way to look at the matter: in terms of old, slow-growing mature sectors and sectors based on newer technologies, experiencing rapid growth. The former include not only textiles, shoes, and the classic nondurable consumer goods, but also steel, machine tools, and, I would now add, automobiles. The more dynamic sectors include the electronics industry, in all its many

TABLE 3
REGIONAL GROWTH SHARES OF MANUFACTURING EMPLOYMENT CHANGE: 1960-1975[a] (numbers in thousands)

	Absolute Growth Increment			Percentage Share of National Growth		
Region	**1960-1965**	**1965-1970**	**1970-1975**	**1960-1965**	**1965-1970**	**1970-1975**
United States total[b]	722.9	2,213.4	–1,467.0	100.0	100.0	–100.0
Northeast region	–110.1	264.9	–936.2	–15.2	12.0	–63.9
North Central region	189.7	624.5	–579.8	26.2	28.2	–39.5
South region	520.6	951.5	23.9	72.0	43.0	1.6
West region	122.7	372.5	25.1	17.0	16.8	1.7

NOTES: a. Employees on nonagricultural payrolls as of March 1 of the respective years.
b. Excludes Hawaii and Alaska.
SOURCE: U.S. Department of Labor, Bureau of Labor Statistics, Employment and Earnings, Washington, D.C.: U.S. Government Printing Office, Monthly. Table 2 was compiled by George Sternlieb and James W. Hughes, "The New Economic Geography of America," Rutgers, New Jersey: Center for Urban Policy Research, January 1977, p. 10.

facets, and certain technologically vital branches of the chemical industry. By and large, the industrial structure of the Northeast and Midwest is heavily weighted by history toward the mature sectors. Those parts of the Northern region linked to the newer sectors have systematically fared better than others. It is in the mature sectors that obsolescent industrial capacity has mainly emerged. In Ohio, for example, 21% of plant and equipment is classified as obsolescent as compared to the national average of 12%; but in the older primary metals and nonelectrical machinery sectors, obsolescence is over 40% (Widner, n.d.:16-17). But the new high technology sectors of the 1950s and 1960s may also be decelerating in the 1970s, a tendency dramatized by IBM's use of its cash surplus in 1977 to purchase its own stock rather than to invest in new directions. The private sector is almost as bemused as federal policy as to where productive investment should be undertaken in the changed context of the fifth Kondratieff upswing.

In the Northeast, a rather remarkable job was done between 1950 and 1970 in supplanting older nondurable industries with durable goods industries often linked to new technologies. A loss of some 338,000 jobs in the former category was just about balanced by 312,000 new jobs in the latter. The rise of federal spending on military and space hardware helped. The Northeast benefitted greatly from its comparative advantage in first class educational, engineering, and research and development institutions. The decline of national security outlays has retarded growth in these sectors. To a lesser degree, however, the comparative advantage of the Northeast in high technology fields remains. The task is to bring this capacity to bear on the resource-related sectors whose importance has risen in the fifth Kondratieff upswing: energy and energy conservation, transport, pollution control, and research and development in general. All this will not be easy. The Northeast confronts a relatively old working force, and, like investment in general, movement is constrained by unresolved national policies for energy and the environment as well as by rather discouraging local tax structures. But if my analysis of the requirements for dealing with the fifth Kondratieff upswing is correct, the sophisticated intellectual and industrial capacity of the Northeast will be greatly needed in the time ahead.

There are also both mature and new sectors in the South and Southwest. Both have generally expanded more rapidly than their Northern counterparts. This happened mainly because of relatively higher rates of both population and income growth. But in certain sectors (e.g., textiles) the South has felt the competition of overseas suppliers. Overall, however, its mix of industries is newer and the problem of obsolescence less acute.

The problems of the Northeast and North Central states were complicated in the 1970s by the rise of the role of services relative to manufacturing in recent years, as Table 2 indicates. We include under services a wide variety of quite different economic activities: wholesale and retail trade; finance, insurance, and real estate; government; education; health services. As nearly as we can measure productivity in services, it is systematically lower, on average, than in manufacturing because it is somewhat less subject to technological innovation. Some services are automatically required, as population, industry, and income expand, and urbanization takes place. Thus, the large expansion of service sectors in the South as well as in the North. Others, notably education, health services, and welfare outlays are subject over a significant range to public policy.

The disproportionate growth of services relative to manufacturing rendered the Northeast and North Central states vulnerable to a situation where the increase of income and population was slowing down and other parts of the country were developing a capacity to generate financial and other services in support of their regional economies. This tendency was exacerbated by the development of a good road transport system in the South and the national shift towards light, high technology industries, less dependent on the dense railroad network which was once a special asset of the North.

Against this background, the coming of the fifth Kondratieff upswing hit hard in the two Northern regions. Somewhat like slow-moving Great Britain in the 1960s, the high-income North had committed itself to enlarged public and private services at a time when its manufacturing base, with high obsolescence in certain sectors, was waning, its rate of population increase was slowing down, and population was actually declining in some major urban areas. The unfavorable shift in the region's terms of trade, as in Britain, reduced real income at just the time it confronted unemployment rates about 2% higher than those in the South and Southwest. The fiscal problems posed for state and local government were only in degree less acute than for New York City. Meanwhile, as noted earlier, the relative rise in energy and agricultural prices accelerated the flow of population to areas producing these products, widened growth rate differentials, and still further weakened the foundations of the economies in the Northeast and North Central states.

As this situation became apparent, the initial reaction of politicians in the North was to seek redress by inducing the federal government to reallocate federal expenditures from South to North. That theme dominated the literature of early 1976 on the subject of regional change. In crude terms the argument was: The rise of the South was a product of disproportionate outlays of the federal government financed by northern taxes; now is the time for the South to contribute disproportionately to the rehabilitation of the North. As experts dug into the data on federal tax and expenditure flows, the evidence suggested a less straightfoward picture of the past and less simple remedies for the future.

For example, when federal tax contributions are calculated as a proportion of income per capita in the states (rather than per capita), the relative contribution of the Northern states to federal revenues is much reduced. On the other hand, when present (inadequate) cost of living indexes are applied to the regions, the North-South real income per capita differential is further narrowed. For example, when the flow of federal grants is systematically correlated with various indicators of economic development in the states, no significant association emerges (Howell, 1976:4). Also, when federal expenditures are broken out by categories and measured in terms of outlays per capita in the various states, the Northeast appears to have been drawing more from Washington than the South in defense contracts, welfare programs, and, marginally, retirement programs (National Journal, 1976). The more rapidly urbanizing South acquired somewhat more for highways and sewers, a great deal more in defense salaries due to the location of military bases. The East and West North Central states fared worse than the South in all categories except highways and sewers; but the West (a high income area) far outstripped all other regions in defense contracts, salaries, and highways, the latter mainly because of the still expanding interstate highway program in the Mountain states.

The analysis and debate about federal tax flows and expenditures can be expected to become more complex but to remain a lively part of the national political scene. My own view is, quite simply, that the major regional changes in the country have been only marginally determined by the balance of federal tax and expenditure flows; and that the future of federal policy and outlays in the regions should be determined by the requirements of the several regions, seen in terms of the nation as a whole. It is a wholesome fact that, as analysis of the nation's regional problems has gone forward in 1976-1977, this is the direction of thought and prescription.

When, for example, the Northeast Governors Conference met in Saratoga in November 1976 the Governors did, indeed, point to the apparent imbalance between tax revenues and federal expenditures in which the South and West profit at the expense of the North. But they went beyond to propose measures which would enlarge investment in energy production and conservation; to rehabilitate the region's transport system; and to expand and modernize industrial capacity in areas of particularly severe unemployment. They proposed manpower training and public works programs, exhibiting considerable sensitivity to assure the latter were undertaken in sectors where investment would prove productive over the long run, e.g., transport rehabilitation, solid waste disposal plants, etc. The conference also considered the complex problem of welfare reform on a national basis, but with an understandable emphasis on the extent to which slow growth, the long prior period of south to north migration, and higher than average unemployment since 1973 have converged to make the welfare problems of the Northeast particularly acute. The institutions and policies, regional and national, do not yet exist to translate these directions of thought into lines of action; but the fundamental issues being explored are significant and hopeful.

In 1976-1977, similar analyses were emerging in the North Central states; for example, the studies of the Academy for Contemporary Problems in Columbus, Ohio. The academy has surveyed extensively the structural changes and problems of the whole East North Central region; and it has made detailed policy recommendations for Ohio. Its policy agenda includes special measures to expand local coal production and to reduce energy consumption; a modernization of the transport system; special incentives to stimulate high growth manufacturing and export industries as well as to rehabilitate aging or obsolete plants. As in the Northeast, there is a call for redirected federal tax revenues, a national welfare plan, and intensified regional cooperation between the public and private sectors to stimulate investment in the directions necessary for further development.

The South and Southwest confront what may appear in the North an easier future; but analysis and policy are also increasingly addressed to its serious structural problems. The excellent report of the Task Force on Southern Rural Development, for example, measures the scale and character of poverty in the South relative to the rest of the country. In 1974 there were still 10.8 million poor Southerners, 13.5 million outside the South. Fifty-four percent of the Southern poor are rural; only 38% are rural outside the South. Evidently, a massive problem of poverty still exists in the rural South roughly matching in scale the more visible urban poverty of the North, but constituting a higher proportion of the total Southern population. The recommendations of the Task Force, notably with respect to retraining and bringing the poor effectively into the working force, are similar to the manpower proposals of the Saratoga program of the Northeast Governors Conference.

As the South and Southwest look to the future, analysts are beginning to perceive a set of investment and policy tasks quite as challenging, in their way, as those in the North and Northeast. For example, the whole irrigated area of the High Plains, from northwest Texas to Nebraska, is endangered by the decline of the underground water basin which supplies it. The region produces a significant part of the American agricultural surplus. Large investments in surface water conservation and transfer will be required to preserve it. In addition, the oil and gas reserves from conventional sources are almost certain to run down over the next generation in Louisiana,

Oklahoma, and Texas. Like the nation, they face a transition to coal, atomic energy, geothermal, solar, and other sources, but they do so with a peculiarly high proportion of their economic structure linked to petrochemicals and other energy-based industries.

On March 31 to April 2, 1977, a conference of the Southwest American Assembly was held in Texas on "Capital Needs of the Southwest: The Next Decade." Its agenda was familiar to analysts in the North, including among the major investment fields: energy and raw materials; plant and equipment for manufacturing; pollution and environmental controls; housing, public facilities, and education.

The challenge to the Sunbelt is, of course, to complete its transition to modernization; to develop further its resources in energy and nurture its agricultural base; and to deal with its social problems, under conditions of rapid population increase, while avoiding to the degree possible the environmental degradation which marked the urbanization of the North. The problem of the North is to bring to bear its enormous potentials in technology, finance, and entrepreneurship; to exploit its energy resources in the new price environment; to rehabilitate its transport system in cost-effective ways; and to modernize the industrial sectors which hold greatest promise for the future. This will require a new regional sense of purpose as well as intimate public-private collaboration.

It is worth asking bluntly: Do the Northeast and industrial Midwest have a future? As we have seen, their present positions are the result of deeply rooted structural problems which decree a greater relative load of obsolescent industries, a heritage of large past commitments to expensive social services, a large concentration of poor in the central cities reflecting a prior north to south population flow over a protracted period. There are those who have concluded that the North, after a century and a half of leadership, should gracefully decline and surrender economic leadership to the South and Southwest. From newspaper accounts, a gathering of economists at Rutgers University in March 1977 developed some such consensus.

I do not share that mood of passive pessimism about the North. For one thing, as the case of Britain illustrates, economic decline is not a graceful process. It is painful, socially contentious, and potentially quite ugly in the political moods and problems it generates. Nations or regions which choose to go down in the style to which they have become accustomed find it a difficult or even tragic path to follow. Moreover, I believe it is unnecessary. Surely, the pattern of economic development in the North will have to change. Surely, the North cannot go on doing what it has been doing if it is to cope with the special pressures of the fifth Kondratieff upswing. Surely, the antiseptic, easy devices of fiscal and monetary policy will not cure the ills of the North. But the lesson of economic development in many parts of the world is that it hinges mainly on the human resources available, mobilized around the right tasks. The North commands both the material and human resources for a great revival.

There are a good many examples of nations which successfully recaptured momentum after falling behind under the weight of mature industries, with substantial obsolescent plant. Post-1945 France and Belgium accomplished such a transformation, as, indeed, did post-World War II New England. The initial assessments of postwar Germany, the economy split in a historically unnatural way, burdened with refugees and great war damage, were exceedingly gloomy. Every northern business, labor, and political leader ought to read Chapter 10 in Jean Monnet's *Memoires.* Its title: "France Modernizes Itself." Monnet describes how France found itself in a state of severe industrial and agricultural obsolescence after World War II and how, through public-private collaboration, led and inspired by a miniscule bureaucracy, it underwent a total technological renovation. Perhaps the two Monnel dicta most relevant from that experience to the situation of the North are: "We do not have a choice. There is no alternative to modernization except decadence"; and "Modernization is not a state of things but a state of mind (esprit)" (Monnet, 1976:306).

Structural transformations are clearly possible if there is a common will to accomplish them, a sense of direction, and a general environment of rapid economic growth. Only those who live in the North can generate the common will, the sense of direction, and the institutions required for common action.

This brief survey of regional change and of thought about future development is, of course, too simple. It cannot be overemphasized that there are great differences among states and, even, within them. There are special problems and prospects in the Mountain states and the far West, Alaska, and Hawaii. But six large general conclusions stand out.

First, a return to sustained full employment and high growth rates on a national basis would greatly ease the special problems of the northern states by simultaneously reducing welfare requirements and expanding tax revenues. Further, a view of the national situation in terms of the regions strongly reinforces the argument I have made elsewhere that sustained full employment is to be achieved by expanded investment in energy and other resource development and conservation fields combined with outlays to reaccelerate the growth of productivity (Rostow, 1977). Without full employment and rapid growth, it will be exceedingly difficult to sustain our welfare services and deal with our poverty problems.

Second, economic as well as social and human considerations require serious and sustained efforts to bring into the working force the large number of Americans now trapped in urban and rural poverty, North and South. The nation should look with equal concern at the problem of the central cities and the impoverished margins of life in the southern countryside. We do not yet command the data to measure the scale of the nation's special investment requirements over, say, the next 10 years in agriculture, energy, raw materials, the environment, the rehabilitation of obsolescent industrial and transport facilities, research and development, manpower retraining. It seems palpable, however, that they will require us to use our manpower resources to the hilt even if the whole automobile sectoral complex is somewhat reduced in size.

Third, the case for a national energy policy which would both enlarge our resources in all the regions and economize their use is greatly heightened as one observes the differential regional impact of the energy crisis which has had us by the throat since the autumn of 1973. Here, with energy policy, we face the most potentially divisive issue in the nation. There is a deep and, in my view, understandable resentment in the southern energy exporting states about some of the attitudes of the North. At one and the same time, the North appears to be demanding both low energy prices while refusing to develop its own energy resources on environmental grounds. This is seen in the South as a straightforward colonial policy of exploitation. President Carter's proposal to the Congress of April 20, 1977, heightened this sense of regional grievance, with its wellhead tax, a politically inspired inadequate price for new gas, and provisions for allocation of gas and oil output back to the wellhead. It would not be difficult to split the nation, yielding an OPEC within it, or a grave, protracted constitutional crisis. The only answer is to have the nation as a whole face up to the fact that energy is and will be expensive;

that conservation and economy are required of us all; and that the energy potentials of the country as a whole—not merely the South and the West—must be promptly and fully exploited under environmental rules of the game that are firmly settled.

Fourth, we require a national welfare policy which would render more uniform the criteria for public assistance; but this may be somewhat more expensive, and will certainly have to be introduced over, say, a three to five year period. And, once again, such a system will be viable and sustainable only in an environment of full employment and rapid growth.

Fifth, within the states and regions new forms of public-private collaboration will be necessary, as well as at the national level. Obviously, there is no way the kinds of problems reviewed in this paper can be dealt with by conventional fiscal and monetary policy. If we are to mitigate or solve our problems, we must deal with particular sectors, regions, cities, and rural areas. We do not have and we do not want a fully planned and directed economy. On the other hand, a public role is inescapable. We have no other course than to learn how to make the public and private sectors work together.

Finally, it seems clear to me, at least, that these problems cannot be dealt with if the states and regions look merely to Washington for salvation. The basic analyses, investment plans, and public-private consortia must be developed within the states and regions. Local capital and entrepreneurship must be mobilized. In the end, however, there is scope for federal assistance in the form of tax incentives, investment capital, manpower retraining programs, and the direction of public service job programs toward areas judged of high priority within states and regions.

Although the creation of state and regional development corporations may have a role, it is likely that a national development bank will be required like the old Reconstruction Finance Corporation. Its authority should extend not only to the fields of energy and energy conservation (where such an institution was proposed to the Congress by the Ford administration) but also to the financing of water development, transport rehabilitation, and other projects judged of high priority national interest. Wherever possible, such a bank should use its authority to guarantee or to marginally subsidize funds raised privately or by state or local governments, rather than engage in full direct financing.

In addition, it would be wise, in the phase ahead, for both state and federal governments to organize their budgets in ways which

CARL A. RUDISILL LIBRARY
LENOIR RHYNE COLLEGE

would separate out authentic investment outlays from conventional expenditures and transfer payments.

There is a warning to be made about any proposal for an increased government role in the investment process: High standards of priority and productivity must be preserved. The experience of Great Britain and other countries with nationalized industries (not proposed here) and government loans to industry suggests that there is an inherent danger of confusing criteria of productivity and simple job maintenance or creation. The latter can lead to increasing public subsidy and the drawing off of scarce investment resources to low productivity tasks. Public authorities tend to persist with lines of investment, even if of low productivity, because they do not face the competition of the market place and because political vested interests build up around such public ventures. If the analysis underlying this essay is roughly correct, the United States and the other industrialized nations have ample opportunities to generate full employment through high-priority and essential investment tasks. None can afford to compound existing tendencies toward declining rates of productivity increase in the advanced industrial world by committing public funds to essentially wasteful enterprises.

NOTES

1. The disparate growth cycles have been extensively documented in Garnick (1977) and by the Bureau of Economic Analysis (1974). The relative price shift also had its effect within regions. For example, the effect of rising energy prices on the relative structure of employment and income within the Sunbelt states was explored extensively at the Conference on the Future of the South's Economy, Boca Raton, Florida, December 12-15, 1976 ("South's Energy States Growing Faster," New York Times, April 2, 1977, p. 25). For data on the disproportionate expansion of personal income in energy related sectors, see W.H. Miernyk (1976a:25, 1976b).

2. On the disproportionate rise of transfer payments, due mainly to high unemployment, as a proportion of income in the Northeast-Great Lakes region, notably in the recession periods of the 1970s, see Renshaw and Friedenberg (1977).

REFERENCES

Bureau of Economic Analysis, U.S. Department of Commerce (1974). Area economic projections, 1990. Washington, D.C.: U.S. Government Printing Office.

GARNICK, D.H. (1977). "The northeast states in the context of the nation." Unpublished manuscript based on a paper originally prepared for the Conference on the Economic Future of the Northeast States, sponsored by the Joint Center for Urban Studies of M.I.T. and Harvard University and the World University of the World Academy of Art and Science, Cambridge, Mass., January 19.

HOWELL, J.M. (1976). "Rebuilding the northeast with the advice of H.L. Mencken: 'For every human problem there is a solution which is simple, neat, and wrong.' " Remarks before the National Conference of State Legislatures, Orlando, Florida, November 29.

MIERNYK, W.H. (1976a). "The changing structure of the southern economy." Paper delivered at the Southern Growth Policies Board Conference on the Future of the South's Economy, Boca Raton, Florida, December 12-15.

——— (1976b). "Regional economic consequences of high energy prices in the United States." Journal of Energy and Development, 1(spring):213-239.

MONNET, J. (1976). Memoires. Paris: Fayard.

National Journal (1976). "Federal spending: The north's loss is the Sunbelt's gain." June 26.

RENSHAW, V., and FRIEDENBERG, H.L. (1977). "Transfer payments: Regional patterns, 1965-1975." Survey of Current Business, May, pp. 15-19.

ROSTOW, W.W. (1975). "Kondratieff, Schumpeter, and Kuznets: Trend periods revisited." Journal of Economic History, 35(4):719-733.

——— (1977). "Caught by Kondrateiff." Wall Street Journal, March 8, p. 20.

——— (forthcoming, a). "Growth rates at different levels of income and stage growth: Reflections on why the poor get richer and rich slow down." Research in Economic History, 3(January).

——— (forthcoming, b). The world economy: History and prospect. Austin: University of Texas Press.

WIDNER, R.R. (n.d.). National shifts and the future of Ohio's economy. Columbus: Academy for Contemporary Problems.

——— (1976). "The future of the industrial midwest: A time for action." Columbus: Academy for Contemporary Problems.

WILLIAMSON, J.G. (1965). "Regional inequality and the process of national development: A description of the patterns." Economic Development and Cultural Change, 13(4):3-84.

Part II

The Political Economy of Sunbelt Cities

Introduction

□ THE SIX ARTICLES IN THIS SECTION cover a wide spectrum of phenomena ranging from the media's characterization of Sunbelt life-styles, politics, and economics to a reconsideration of the "power shift" hypothesis in light of the rising prominence of multinational corporations. In presenting these articles, we hope that they will illuminate some of the factors contributing to the rapid economic transformation of this region as well as highlight some of the unique features which distinguish the sociopolitical processes in the Northeast from those which are typical of the Sunbelt.

There appears to be little dispute that Houston is now the "shining buckle" of the Sunbelt. But this was not always true. According to William D. Angel, Jr., as recently as the turn of the century, most Texans would have assumed that Galveston, blessed with a fine natural harbor, was destined to emerge as the premier city along the Texas Gulf Coast, while Houston, situated nearly 60 miles inland beside the shallow Buffalo Bayou, would be relegated to relative obscurity. Angel asserts that the cause of this sudden and dramatic reversal of fortune for Houston is that the quality of local entrepreneurship is a more important developmental resource than natural location factors, and thus, Houston's apparent natural disadvantage was far from fatal. Because Houston was blessed with an abundance of farsighted businessmen who constantly sought to develop new innovations and who were not content to merely emulate previous business practices, her economy surged forward to overshadow Galveston as well as every other city in Texas. Specifically, Angel claims that the entrepreneurial activities required to construct the Houston ship channel and to bring the Manned Spacecraft Center to Houston are the dominant factors contributing to the growth and development of that city. Both of these episodes involved path breaking innovative behavior and thus qualify as further examples of the city building entrepreneurial spirit.

In view of all the hoopla and publicity which the South received after the election of President Carter, it is easy to forget that only 10 years ago, this region was the object of the media's scorn and derision. Gene Burd analyzes this abrupt about-face and concludes that the media has always presented an unbalanced and generally distorted portrayal of the Sunbelt. In the beginning, the dominant journalistic image was that of a racist, conservative region whose qualities were best summarized by H.L. Mencken's now famous sobriquet, "The Bible Belt." The vivid impression of the civil rights struggles fought against racist, redneck Southerners armed with high pressure fire hoses and cattle prods did little to efface this image. Suddenly, however, negative characterizations ceased as the nation discovered the good life of ceaseless luxury and endless summers which were open to all residents of the Sunbelt. Unqualified superlatives replaced negative reporting as the nation's journalistic establishment jumped on the booster bandwagon. Any negative imagery or caveats were buried in innocuous asides relegated to the end of the media's paeans. Thus, according to Burd, "backward becomes rustic, ignorance becomes simple and uncomplicated, and reactionary becomes old fashioned" in the current media bliss blitz of the Sunbelt.

The first two articles in this section are concerned with the role of urban boosters as agents promoting the growth of Sunbelt cities. Arnold Fleischmann further elaborates upon this theme by analyzing some of the political ramifications of the booster phenomenon. Specifically, the process of urban annexation is often offered as an explanation for the disparate growth trends observed in the Northeast and the Sunbelt. While northeastern cities are invariably hemmed in by hostile suburbs which vehemently resist any annexation overtures, many Sunbelt cities are favored with lenient annexation laws that permit them to recapture and reabsorb any would be suburbanites. But to successfully pursue an active annexation strategy requires a strong political coalition of builders, real estate speculators, and politicians. These are the urban boosters which are the subject of Fleischmann's research. His study details the links between annexation and political structure and documents three distinct stages in San Antonio's postwar annexation history.

Currently, Fleischmann reports, the pro-growth coalition which encouraged San Antonio's spatial expansion has suffered several political defeats. Whether San Antonio will continue to grow in the absence of political control by these booster elements remains to be seen.

The next two selections focus on the deleterious effects engendered by rapid growth. The emphasis moves from the process of growth to the impacts and distributional consequences accompanying rapid development.

Peter A. Lupsha and William J. Siembieda strongly suggest that we should not expect convergence in the interregional levels of public services despite the rapid economic expansion in the Sunbelt. Although the rising affluence will provide the fiscal resources for improved service delivery systems, the differences will persist due to the variances in cultural and political perspectives toward public functions and responsibilities. In the Sunbelt, for instance, the elected political elites have traditionally believed that the provision of many common services—public transportation, emergency medical services, paved streets, curbing, etc.—went beyond the legitimate obligations and functions of the polity. In addition, Lupsha and Siembieda contend that there must be some agreement on which equity model is to be espoused. In the Northern metropolitan areas, the traditional market equity model was replaced by one which favored equality of opportunity and increasingly, equality of result. These cultural factors favored a more active role for state and local governments. In the Sunbelt, however, the espousal of a philosophy of privation coupled with political structures that are designed to mute the demands of the economic underclasses encourages a minimalist conception of government responsibilities. Thus, according to Lupsha and Siembieda, continued divergence, not rapidly approaching convergence, will typify the interregional patterns of public services.

Robert E. Firestine also analyzes the distributional impacts of rapid growth in the Sunbelt but focuses on the emerging patterns of intrametropolitan disparities. He suggests that within the Sunbelt metropolis, personal income inequality may well increase in the near future, regardless of the expected continuation of economic growth. Concurrently, in-migration of higher income, upwardly mobile households will reinforce the outward relocation of jobs and residences from many Sunbelt central cities. Taken together, these events will likely worsen the economic plight of the least advantaged citizens of the Sunbelt metropolis while sowing the seeds of future fiscal conflict between their central cities and their surrounding suburbs. Thus, according to Firestine, the central city suburban disparities which have characterized Northeastern metropolitan expansion are being rapidly replicated in the Sunbelt. Unless these trends are reversed, many of the same problems which plague the Northeast, such as white flight, an eroding tax base, and an antipathy to annexation, will soon be haunting the Sunbelt.

The last article by Robert Cohen addresses the question of a regional power shift. Kirkpatrick Sale unequivocably asserts that the locus of economic power has truly shifted from the Northeast to the Sunbelt. Cohen is more dubious. He claims that as the modern corporation developed and transformed itself into a multinational entity, it sought new markets and attractive conditions for production. Like many developing nations, the Southern Rim offered both. Numerous factories moved south and many foreign firms invested there. Thus, from a corporate power standpoint, the rise of the Sunbelt, like the development of a third world nation, can be viewed as a new region's integration into the world as well as the national economy. And like a dependent nation, much of the Sunbelt's industry and a significant portion of its finances remain under the control of outside economic actors. As Cohen demonstrates, the corporations headquartered in the Southern Rim have not been particularly active overseas nor have they been major participants in the nation's research and development effort. The linkages between banks in the Sunbelt and local corporations are weak, and the region's banking and corporate service sector, with the exception of activities generated in San Francisco, do not compete with those in Chicago and New York. Thus, while there is no disputing the fact that the Sunbelt is presently expanding more rapidly than the Northeast, there is no evidence to suggest that this has instigated a massive power shift from the Yankee empire to the Cowboy economy. With very few exceptions, the bulk of the corporate decision making and power still resides in the Northeast.

4

To Make a City: Entrepreneurship on the Sunbelt Frontier

WILLIAM D. ANGEL, Jr.

□ URBAN DEVELOPMENT is often associated with a favorable constellation of location variables such as access to inland water systems, proximity to valuable natural resources, or the excellence of the local port facilities. While there is no doubt that urban development will be aided in those areas where location considerations are naturally favorable, these conditions are not sufficient or even necessary to guarantee a city's economic success. They will not preordain urban growth because nature does not make cities; people make them.

Several years ago Seth Low remarked, "The problem in America has been to make a great city in a few years out of nothing." His statement is not a misguided attempt to apply the theory of spontaneous generation to urban affairs. Rather, it reflects the realization that cities had to be meticulously constructed, often in the face of seemingly insurmountable natural obstacles.

Throughout history a unique type of capitalist—the urban entrepreneur—has played the leading role in surmounting these natural barriers and converting them into a basis for dynamic growth. Thus, the "taming" of the frontier can be viewed, in a very real way, as the process whereby entrepreneurs developed cities by breaking through the barriers erected by the seemingly impenetrable wilderness. Today entrepreneurs are still at work, struggling to preserve the vitality of the old urban areas and to develop new cities in the Sunbelt. This paper will highlight the city building role of urban entrepreneurs and special emphasis will be placed on the entrepreneurial function supporting Sunbelt development.

THE ENTREPRENEURIAL CONCEPT

Urban historians have accorded special recognition to the role of private initiative in fomenting the growth of cities. For instance, Sam Bass Warner (1968) has described the American city as the "Private City." He argues that when capitalists were successful, the city was successful. But despite this recognition of the prominent role of individual initiative, there have been few attempts to distinguish entrepreneurial functions from nonentrepreneurial ones. Part of this failure is due to the fact that numerous individuals participate in local economies and urban scholars tend to describe their behavior in catch-all terms. Such concepts as "businessman," "capitalist" or "entrepreneur" are employed interchangeably. Unfortunately, this imprecise conceptualization only serves to cloud any explanatory power which these terms might possess. As a result, urban historians "feel" the importance of various business personalities without really understanding their impact. In order to remedy this situation, we must first define the concept "entrepreneur" so that the precise role of different elements within the business sector can be appreciated.

Entrepreneurs are innovative capitalists. They engage in legally prescribed business endeavors and use society's resources in innovative ways to attain profits.[1] For them, innovation is a means to an end and not an end in itself. They differ from nonentrepreneurs by their reliance on innovation as their modus operandi. Nonentrepreneurs are capitalists who seek only to accumulate capital by mimicking established business methods. They do not seek change in the mode of production, nor do they try to attain greater profits by using innovative techniques. Like entrepreneurs, they are motivated by the profit impulse, but unlike entrepreneurs, nonentrepreneurs will decline to use innovative procedures until someone else has adopted such methods with profitable results.

The American entrepreneur has been a "barrier buster." The foresight and creativity of his innovativeness is informed by his personal vision of economic growth and accumulation. He pushes himself to the limits of accumulation potential and goes beyond what others view as practicable business. For the entrepreneur, the parameters of economic growth are not nearly as well defined as they are for others. Once he penetrates these phenomenological barriers,[2] new accumulation possibilities open, and other businessmen are quick to copy the newly developed business opportunities. Matthew Josephson (1934:289) has perhaps best described the nature of barrier busting when he wrote that the entrepreneur is "one who feels the turn of the current before others; at first he seems to be navigating against it; then it shifts as foreseen, and he moves forward with an overwhelming force behind him." These countervailing currents may be concrete and physical (e.g. mountains, primeval wilderness, or uninhabited prairie), and yet they may also involve abstractions (e.g. developing new markets or finding ways to defeat economic rivals). However, as Josephson infers, the significance of entrepreneurship lies in overcoming those currents generated by obstructions to the quest for capital.

It is also important to recognize that entrepreneurship can be an individual effort or a group initiative. In many cases innovation implementation may require the combined efforts of several entrepreneurs. Altering a city's infrastructure to make it more economically attractive may necessitate community entrepreneurial initiatives, rather than piecemeal, individual entrepreneurial acts. In this case, the capitalists who begin and direct the innovative efforts are entrepreneurs.

THE ENTREPRENEURIAL IMPULSE

Urban entrepreneurs were crucial in the penetration of the Western frontier. The frontier was a potential resource which could bring immense wealth to those individuals who were capable of tapping it, but, at the same time, its primitiveness constituted a barrier which had to be overcome before that wealth could be realized. In Rosa Luxemburg's (1951:366) terminology, the frontier was a natural economy, whose destruction was a crucial prerequisite for capital accumulation. In their struggle against natural economies, urban entrepreneurs substituted a market economy for a self-sufficient one. The frontier areas were integrated into the commercial network as farmers began raising cash crops in order to purchase consumer goods. Therefore, Turner (1956:1) was not incorrect in asserting: "The existence of an area of free land, its continuous recession, and the advancement of American settlement westward explain American development." Indeed, the frontier was important, but its significance lay in the resource base and consumer markets it provided for the developing capitalist economy.

Urban entrepreneurship centered around this urban-hinterland struggle. If frontier settlements were to succeed, they had to acquire access to both frontier resources and consumer markets. Accomplishing this was contingent upon the entrepreneurial character of the city's leading businessmen. Where entrepreneurs flourished and where those entrepreneurs capably "interposed" their community between hinterland and market, the community would experience economic success.

Urban entrepreneurship on the frontier should be understood in this context. Capitalists in one community often competed against capitalists from another for resources and markets. One community's success usually spelled the other's relative failure. Entrepreneurship involved devising ways in which one settlement could acquire greater relative access to hinterland resources and consumer markets and thus capture the bulk of the trade in an economic region. Through the use of innovative methods, entrepreneurs burst this very important barrier, and successful "barrier busting" also ensured future economic success for the city.

Natural locational features were almost totally irrelevant in this process. Acquiring access to resources and markets sometimes involved changing the natural trade flows dictated by topography. Where rivers were shallow, urban entrepreneurs would sponsor dredging and other improvements. If prevailing water networks dictated north-south trade patterns, a city's entrepreneurs would support railroad or canal building to redirect the trade flows toward their city. Thus, entrepreneurship implied breaking the barrier associated with the "natural inaccessibility" of hinterland regions. In breaking these barriers, entrepreneurs built cities on the frontier.

For example, the construction of the Erie Canal in 1825 enabled New York City entrepreneurs to acquire a tremendous commercial advantage over mercantile capitalists in Baltimore and Philadelphia.

Prior to 1825 the latter cities had been much closer to hinterland resources than was New York, and as a result most of the hinterland trade flowed through Baltimore and Philadelphia rather than New York. But New York entrepreneurs urged the state to build a canal from the Hudson River to Lake Erie and thus provide the city with a direct water route to western agrarian resources. The Erie Canal complemented other innovative activities that New York City entrepreneurs had earlier initiated: commencement of regular packet service between the U.S. and Europe, and the ability of New York merchants to control the South's cotton market.[3] By 1825, the city's merchant community enjoyed a tremendous advantage over economic rivals in competing sites. This advantage was not due to natural endowments but resulted from the efforts of the city's merchant entrepreneurs (Albion, 1939:38-54, 76-94).

New York City's success, at the same time, constituted a barrier to latent entrepreneurs in Baltimore and Philadelphia. Between 1825 and 1860, Baltimore and Philadelphia capitalists tried to overcome the commercial advantages attained by New York City. Philadelphians pursued the construction of a "mongrel" transportation system consisting of a mixture of rail and canal lines. This Pennsylvania Mainline connected Philadelphia with Pittsburgh and provided the city with improved access to the hinterland. Baltimore entrepreneurs, on the other hand, chose to develop rail links to the West in the hope of undercutting both New York's and Philadelphia's commercial superiority (Rubin, 1961:6-14). These separate responses present a contrasting and perplexing case study of urban entrepreneurship, the penetration of the frontier, and the growth of urban economies.

Philadelphia's capitalists do not fit the entrepreneurial pattern. Although they were clearly instrumental in constructing Pennsylvania's system of public works, their arguments on behalf of a canal-oriented Mainline indicate a lack of innovation and creative foresight, two barrier busting essentials. Philadelphians regarded the railroad primarily as an auxiliary that would be useful only where geographical conditions would prevent canal construction (Rubin, 1961:26). They also argued that the canal had proved its utility in both England and the United States, whereas the railroad had not. Building a railroad was considered too risky, while canal construction was thought to lead to sure success. Philadelphia's merchant capitalists were not innovators; rather, they were only interested in imitating those entrepreneurs who had previously pioneered in canal building (Rubin, 1961:38).

Baltimore's capitalists clearly were more entrepreneurial than were Philadelphia's. Baltimore entrepreneurs were not as fearful of the uncertainty of railroad construction. They were less impressed with arguments concerning the superiority of canals than were Philadelphians (Rubin, 1961:72). Also, unlike Philadelphia capitalists who relied on state funds to construct transportation systems, Baltimoreans willingly invested their own capital in railroad enterprises, thus leading by example rather than relying on any form of state subsidy or investment (Rubin, 1961:75). The city established a governmental committee to implement local transportation policy, and, in 1836, it heeded the advice of local entrepreneurs and purchased $3,000,000 of stock in the Baltimore and Ohio Railroad and $1,000,000 of stock in the Baltimore and Susquehanna Railroad (Goodrich and Segal, 1953:17-19).

Despite the greater entrepreneurial prowess of Baltimore capitalists, Baltimore assumed a secondary position to Philadelphia by the late 19th century. Part of Baltimore's later problems may be traced to the city's rivalry with Philadelphia for Susquehanna Valley commerce. During the 1830s Baltimoreans initiated a series of railroad building efforts in the hopes of undercutting Philadelphia's commercial access to the Susquehanna Valley (Livingood, 1947:116-130). The resulting Baltimore and Susquehanna System (later the Northern Central) failed to generate sufficient revenues for profitable operation and was taken over by the Pennsylvania Railroad company. As a result Philadelphia consolidated its hold on Susquehanna trade.

The difficulty was not entirely of Baltimore's own making. Officials of the Baltimore and Ohio Railroad viewed the Northern Central as a threat to their commercial interests and commenced a rate war which the Northern Central could not survive (Livingood, 1947:140). By the 1850s, Baltimore's entrepreneurs had little control over rate policy even though the city had substantially invested in rail companies. Private director-investors in the Baltimore and Ohio began to operate the line in light of what would be most profitable for them and did not consider what would be best for Baltimore's economy (Goodrich and Segal, 1953:23-25).

Also, Baltimore's rail links to the West had to traverse either Pennsylvania or Virginia. Both states were hostile to the idea of authorizing any transportation system that would give Baltimore a substantial advantage. For example, the Virginia legislature delayed attempts to charter the B & O, and it was not until 1853 that the Baltimore and Ohio System finally reached into Ohio. Meanwhile, Pennsylvania had replaced the mongrel Mainline with a complete rail system and thus subverted another Baltimore attempt to develop a commercial advantage over Philadelphia (Albion, 1939:382-384).

But Philadelphia did not really succeed either. Pennsylvania's Mainline was never a serious competitor for New York's Erie Canal. Shipping commodities between Pittsburgh and Philadelphia was expensive and time consuming. Numerous transshipments between rail and canal lines were required, and the system possessed a limited capacity to handle an extensive amount of trade (Rubin, 1961:16-17). One can only wonder what would have occurred had a complete rail system been adopted much earlier. Rails would have offered much faster transport than the Erie Canal, and rates would have been competitive. While the Erie Canal was frozen during the winter, a Philadelphia oriented rail connection to the West would have operated year round and might have successfully diverted the trade of western farmers away from New York. Such an argument is only speculative, but it is clear that Philadelphia's nonentrepreneurial decision to pursue the mongrel Mainline strategy did not provide the city with a commercial advantage relative to New York City.

It is interesting to note that in the 20th century, the basic fabric of those "barriers to accumulation" has changed little since the days of the urban frontier. These barriers are perhaps less physically imposing than they once were and are more related to those national economic conditions characteristic of a mature capitalist economy. Their effect, however, is still the same. Today's barriers to accumulation are not so much "nature-made" as they are "market-made." This involves perception of market constraints which inhibit local businesses and industries from maximizing profits. The 20th century entrepreneur attacks these barriers, like entrepreneurs of the past, with the intent of expanding his firm's profit making capabilities.

Hence, it may only *seem* that entrepreneurial skill was more relevant in the 19th century than it is in the 20th. Some scholars argue that locational factors—not entrepreneurship—are the primary determinants of industrial locational decisions and of governmental decisions to award prime contracts. In a similar vein, these location theorists point out that firms will locate in communities where production and distribution costs are lowest. Such factors as labor costs, energy supplies, critical resources, inter and intraurban transportation systems, and other external economies are likely to be viewed as more important in corporate locational decisions than is entrepreneurship (Duncan et al., 1960:23-81; Fuchs, 1962; Heberle, 1954; Duncan and Lieberson, 1970).

Certainly, locational factors are important in contemporary urban development. Local entrepreneurs, however, still have some influence upon various public and private site selection committees. They may convince prospective firms of the community's ability to mitigate any constraining feature, or they may advertise the relative importance of various locational advantages which already exist. It may well be that a firm or contracting agency has several equally desirable sites from which to choose. In such circumstances, entrepreneurs mobilize community support, develop booster strategies, and create the requisite infrastructure; all of which are required to attract the prospective business.

It is also important to recognize, as do Duncan and Lieberson (1970:20), that "what a city does depends not only upon where it is, but upon what it has done in the past." That is, what a city has done in the past involves the notion of providing infrastructure which will serve future economic development. Entrepreneurs can be important in this respect as developers of infrastructural assets which will permit the city to overcome any natural "disadvantages."

SUNBELT ENTREPRENEURSHIP

Entrepreneurship in the Sunbelt is analogous to the type of entrepreneurship practiced during the penetration of the frontier. In the 19th and early 20th centuries the South's industrial structure was agrarian and not industrial. The region produced cotton, sugar, tobacco, lumber, cattle, and rice in great quantities, while its urban centers were primarily commercial nodes designed to serve the agrarian hinterland. Therefore, entrepreneurial activities in Southern and Southwestern cities first centered around establishing the communities as viable service points for the rural economy. Accordingly, local entrepreneurs actively solicited the establishment of railroad facilities and processing plants designed to serve farmers and ranchers in the surrounding territory.

Sunbelt entrepreneurs created a flexible and resilient local infrastructure, which also ensured their city's future success. These infrastructural assets not only had to serve agrarian market requirements but also had to be sufficiently flexible to attract so called "leading edge" economic activities. Admittedly, the entrepreneurial intent was to develop infrastructure for the exploitation of hinterland markets. But the ultimate impact was the creation of infrastructural advantages capable of serving a variety of economic sectors. When the country's industrial and tertiary sectors became more prominent, many Sunbelt cities were in a position to capture a share of that activity. Such cities became less important for their primary sector functions and oriented themselves toward serving 20th century economic requirements.

A prime example of such a city is Houston, Texas. Certainly Houston has possessed certain natural features which have served it well: location in the rich agricultural region, proximity to the Gulf of Mexico, and the presence of an oil-laden hinterland. Although Houston's natural endowments may seem overwhelming, one must also recall that during its early history, the city was overshadowed by its seaport rival–Galveston. Houston–located along the shallow, vegetation infested Buffalo Bayou–was approximately 50 miles from the gulf, but Galveston was located on a large island in the only natural harbor along the Texas Gulf Coast. This seemed to ensure Galveston's destiny as a major shipping center for Texas's cotton, rice, and sugar.

Entrepreneurs in Houston were not content to remain in second place. As early as 1849, Houstonians indicated that, while Galveston might become an important commercial center due to its "vantage on the Bay," it was also isolated and difficult to reach from the mainland. They speculated that if an inland port could be established, it would soon surpass Galveston (Houston Democratic Telegraph and Texas Register, December 13, 1849).

The first step to establish Houston as that inland port involved making the city a rail hub for Texas. If this could be accomplished, cotton, sugar, rice, and other staples could be funnelled to Houston along rails and then shipped to Galveston for overseas transport. The impetus for creating this inland port is mainly due to private initiatives on the part of Houston entrepreneurs. Houston entrepreneurs intially organized and directed the railroads that radiated from the city (e.g. the Houston Tap and Brazoria, the Houston and Texas Central, and the Texas and New Orleans). They lobbied the Texas legislature for loans and land grants and thus received the initial investment capital for construction. They also successfully established local support in communities along proposed routes by convincing their citizens to subscribe to stocks in Houston-directed rail companies.

By 1860 Houston had become the principal Texas rail center, with Galveston remaining the state's primary port. Businessmen from Galveston were perfectly willing to live with the arrangement whereby Houston would serve as a collection point for agricultural produce and Galveston would function as a shipping point for oceangoing transport. Houstonians were not. In 1856, at a time when Houston was just beginning its railroad expansion, the Houston Telegraph (November 18, 1856) declared, "With good compresses, cotton might even be put on board vessels in Galveston Bay without touching the island at all." Galveston merchants scoffed at this and thought the idea that "Houston could do her own shipping from the mouth of the Buffalo Bayou [was] a little too hard to swallow" (Galveston Weekly News, November 25, 1856).

The principal barrier facing Houston's entrepreneurs was the city's lack of access to the gulf. However, Galveston too was not fully accessible to deepwater shipping since the port was obstructed by sandbars. Shipments from both cities had to be first lighted on barges and then shipped to vessels anchored in Galveston Bay. Galveston Harbor, however, was thought to be "most susceptible to such improvements as would constitute it a harbor of the first class" (U.S. Bureau of Statistics, 1884:3). Galveston achieved federal support for the harbor project, and, by 1900, the sandbars had been sufficiently "scoured" to allow deepwater ships to load and unload directly at Galveston's wharves (U.S. Army Corps of Engineers, 1901:462).

Although the Buffalo Bayou had also been deepened and widened, through a mixture of private and public efforts, by 1909 it was only 18 feet deep and could not accommodate deepwater shipping. Without deepwater facilities, Houston's entrepreneurs contined to experience substantial losses of trade to the Galveston port. During the early 1900s, working through congressional representatives, they received small amounts of federal money to support the deepwater project. Piecemeal support, however, would not do the job, since by 1909 it was estimated that almost $4,000,000 would be needed to fulfill the Houston entrepreneurs' deepwater ambitions (Houston Post, January 19, 1909).

On January 18, 1909, a group of Houston entrepreneurs met in Mayor H.B. Rice's office to determine a strategy for making Houston a deepwater port. Among those present were: Rice (also president, Suburban Homestead Company and vice-president, Merchants and Planters Oil Company); Thomas H. Ball (private attorney and U.S. Representative from Houston, 1897-1903); S.F. Carter (president, Lumberman's National Bank); A.S. Cleveland (secretary-treasurer, Houston Compress and Warehouse Company and vice-president, Houston Business League); H.W. Garrow (vice-president, Inman Cotton Compress Company); C.P. Dillingham (president, South Texas National Bank); R.M. Johnston (editor, Houston Post) Frank Bonner (vice-president, Kirby Lumber Company); A.L. Nelms (president, Houston Cotton Exchange); City Attorney W.H. Wilson; Harris County Judge A.E. Amerman; and several local politicos (Ball, 1936; Houston Post, January 19, 1909; Houston City Directory, 1910-1911). At this meeting it was agreed that the federal government would have to be engaged in a much more substantial fashion if the project was to succeed. They proposed to split the financing of the four million dollar project down the middle with Houston raising $2 million and the federal government "appropriating a like sum" (Houston Post, January 19, 1909).

This represented, by far, the largest direct financial grant the federal government had ever made to a local government and the barriers blocking successful implementation of this strategy were numerous. The local matching monies could be acquired only through bond sales, but the city had no legal authority to issue bonds for improvements outside the city limits. Nor did Houston's charter allow the city to consign its revenues to a higher governmental authority. Thus, at the January 18th meeting, it was also decided that these problems could be solved by creating a separate governmental organization for the sole purpose of issuing bonds to finance navigation improvements. Since Texas law did not authorize the formation of these governmental units, Houston entrepreneurs had to persuade the state legislature to enact enabling legislation. Accordingly, local attorneys W.H. Wilson and T.H. Stone were dispatched to Austin to ensure successful passage of the measure (Houston Post, January 26, 1909). Their lobbying efforts were successful, and in February the Texas legislature passed a law authorizing the creation of "navigation districts to . . . construct and maintain canals and waterways." These navigation districts would also have the power to issue bonds and levy taxes (Houston Post, February 16, 1909; Texas Laws, 1909:32-45).

With the legal snarls temporarily solved, Houston entrepreneurs still had to convince Congress to go along with the matching fund ploy. In December, 1909, a delegation of Houston entrepreneurs and politicians appeared before the House Rivers and Harbors Committee. Ball, the delegation's spokesperson, reportedly impressed upon the committee that this was not just another "pork barrel project" and that Houstonians were so anxious to get the project underway that they were "willing to tax themselves to provide for the great facilities for enlarged commerce that it would give" (Houston Post, December 11, 1909). Ike Standifer, a Houston attorney, remained in Washington to oversee appropriations proceedings, and in June, 1910, a ship channel appropriation for $1,250,000 passed both houses. The funds, however, could be issued only when the local navigation district provided matching funds (Houston Post, June 10, 1910).

With Federal monies now available, a navigation district had to be created and bonds issued so that the congressional appropriations could be received. W.D. Cleveland (president of the Houston Compress and Warehouse Company) organized the effort to win voter approval for both the navigation district and the bond issue. On June 7, 1910, at a meeting of Houston businessmen and politicians, it was agreed that the navigation district should include all of Harris County and that an extensive public relations campaign should be mounted to win popular support for the project. The latter was especially crucial since approval by two-thirds of Harris County's property owning voters was required (Houston Post, June 8, 1910). Cleveland then appointed an executive committee to carry out the "Deepwater Campaign."[4]

Although the Deepwater Campaign began with great enthusiasm, the effort was not without its problems. By August, 1910, attendance at rallies and promotional excursions had fallen off. There is also some indication that Houston's middle- and working-class property owners were unwilling to see the Navigation District (and its taxing power) include all of Harris County. They resented the fact that neighborhood improvements had long been neglected and that they were now being asked to approve the creation of a taxing authority which had little to do with their immediate needs. Thus, members of Houston's North End Improvement Association—an organization composed of middle- and working-class residents of Houston's Fifth Ward—indicated that they would support the Harris

County Navigation District only if a bond issue election for a proposed Main Street Viaduct was held at the same time[5] (Houston Post, August 5, 1910). A Business League spokesman later responded, "As it stands now, they [the voters in the Fifth Ward] want the viaduct and will vote for the channel to get the viaduct" (Houston Post, August 6, 1910). On September 1, representatives from the Deepwater Campaign Executive Committee and the North End Improvement Association met in Mayor Rice's office. They agreed that elections for both issues should be held on the same day, and each group promised to work vigorously on behalf of the other's interests (Houston Post, September 2, 1910). After another public relations spurt, the election was finally set for January 10, 1911, at which time both the Ship Channel and the Viaduct issues received 95% voter approval (Houston Post, January 11, 1911).

But the Navigation District Board (the governing body of the Navigation District) could not sell the $1,250,000 bonds that the voters had approved. Interest rates on the bonds were a low 4.5%, while other securities were offering more attractive yields. This problem was exacerbated by the fact that navigation bonds were a new form of security, and investors were hesitant to tie up their capital in them. Thus, by fall 1911, with most of the bonds still unsold, several local bankers met with C.G. Pillot (chairman of the Navigation District Board) and County Judge A.E. Amerman and agreed to purchase the bonds (Houston Post, October 30, 1911). When this deal was consummated, the participating banks received deposits to the credit of the Navigation District equal to the amount of bonds each purchased (Houston Post, March 29, 1912). With the bonds sold, the federal government finally released the $1,250,000 in matching funds, and dredging operations commenced in June, 1912. In September, 1914, the Houston Ship Channel finally opened (Houston Post, September 7, 1914).

The Ship Channel inaugurated new commercial and industrial vistas, especially in the newly developing petrochemical industry. It is interesting to note that although Houston had extensive oil fields nearby at the time the Ship Channel was begun, most of the oil refining in Texas was located in Beaumont and Port Arthur (Rister, 1949:78-79). Once the dream of deepwater at Houston became a reality, oil companies began showing interest in locating their operations in Houston. As early as 1910, J.S. Cullinan, president of the Texas Oil Company, began organizing a move to acquire channel frontage for his oil refining operations. In November, 1910, he led a delegation of Northeastern and European capitalists on a promotional tour of the proposed Ship Channel. At the time, he claimed that he "was anxious to show the visitors what Houston has" (Houston Post, November 20, 1910). After the ship channel became fully operational, numerous other oil companies began to follow his lead and located refineries along the channel (Rister, 1949:229-230).

Entrepreneurial initiatives stand out in this story. Important innovative behavior was exhibited by Houston capitalists such as those who met in Mayor Rice's office in January, 1909, and those who led the 1910-1911 Deepwater Campaign. The creation of the Navigation District and the matching fund gambit were obvious innovations that reached fruition in the Deepwater drive. The barriers were numerous: getting congressional appropriations, convincing voters of the project's necessity, and, finally, selling the bonds themselves. In all of these, the barrier busting role of Houston's entrepreneurs was preeminent.

The immediate entrepreneurial goal was to transform Houston into an agrarian market center. Bringing deepwater facilities to the city established the local infrastructural advantage necessary to achieve this goal. It meant that "over 4 million bales of cotton from Oklahoma and Texas would find its way to the sea through Houston, where the grain of the West . . . would be pouring its treasure" (Houston Post, June 8, 1910). But the Ship Channel also provided an infrastructural asset that would attract mid-20th century industrial concerns. Certainly, the Ship Channel did expand Houston's agricultural market capabilities; by 1920, it emerged as the largest inland cotton market in the world, surpassing Galveston as Texas's leading cotton market. As the city adjusted to industrial pursuits, its agrarian interests began to wane. The city failed to provide enough of the specialized services required exclusively for cotton trade, and by 1947, Galveston reemerged as Texas's leading cotton center (Boehm, 1975:17, 19, 30). This was of little import, however, since Houston had become economically diversified and more capable of attracting leading edge businesses and industries. Galveston was no longer a serious competitor.

It is also interesting to note the importance of linkages between entrepreneurs and local, state, and national politicians. Houston entrepreneurs were not as visible in these activities as were the attorneys and politicians who carried out the Deepwater Campaign.

Ball, of course, stands out as a prime example. J.F. "Jake" Wolters, local attorney and later a U.S. senatorial candidate, led the final push for voter approval of the Deepwater proposal. Other politicians prominent in the campaign were County Judge A.E. Amerman, Sinclair Taliferro (local attorney), Joseph Hutcheson (former U.S. Representative). J.Z. Gaston (City Finance Commissioner) and, of course, Mayor Rice.

These political entrepreneurial relationships have remained an important part of the Houston political economy. Close relationships between politicians and local entrepreneurs have served to protect and expand the Houston economy, just as they did when Houstonians struggled to develop the Ship Channel. No example illustrates this point better than the case of Houston's acquisition of the Manned Spacecraft Center.

In 1961 the National Aeronautics and Space Administration was looking for a site for a Manned Spacecraft Center to design and test spacecraft for the Apollo moon missions. Certainly Houston possessed certain locational features that made it attractive: deepwater facilities capable of handling the bulky lunar equipment, educational institutions, extensive industrial facilities, and a technologically competent labor market (Houston, September 1962, pp. 30-37). However, I would not go so far as to claim that these locational advantages were more important than the political-economic relationships which (might have) influenced NASA's decision (Oates, 1964:355). Many of the advantageous features clearly were established by entrepreneurial initiatives. For example, had it not been for the successful Deepwater Campaign and the subsequent industrial development fostered by the Ship Channel, Houston might not have been nearly as attractive as an MSC site.

The MSC site selection process was dotted with a series of political and economic intrigues. Of prime importance was Vice-President Johnson, who also served as chairman of the National Aeronautics and Space Council. According to Albert Thomas (one of Houston's congressmen), Johnson "did plenty with a capital P" to bring the MSC to Houston (Houston Post, September 24, 1961). Johnson, along with Thomas, was said to have argued "persuasively . . . on behalf of Houston's advantages" (Houston Post, September 20, 1961).

Albert Thomas himself played a key role. He chaired the House Independent Offices Appropriations Subcommittee, the one responsible for approving NASA funds. Although Thomas described himself as being "just the water boy," he also stated somewhat cryptically, "The key to the selection [of the MSC] seems to lie in Congressional approval of the vastly increased budget for Space asked by this administration" (Houston Post, June 13, 1961). And once Houston was selected as the MSC site, Thomas indicated to President Kennedy that he would "support the President on certain matters before his subcommittee" (U.S. National Aeronautics and Space Administration, 1969, I-33).

Clearly Houston had friends in Washington.[6] But local connections were important, too. During the summer of 1961, Morgan Davis, president of Humble Oil Company, donated a 1,000 acre tract of land to Rice University, with the understanding that Rice would offer the land gratis to NASA for the MSC (Houston Post, September 20, 1961). The property was 22 miles southwest of Houston and said to be worth $3,500 per acre. It was only lightly populated, having been set aside for Humble Oil exploration (Houston Post, September 20, 1961; Houston, March 1964, p. 23).

George R. Brown was another important local entrepreneur. He served (and still serves) as chairman of the board of Brown and Root, an important Houston architectural and engineering firm. At the time Humble Oil donated the 1,000 acre NASA "bait" to Rice University, Brown chaired Rice's board of trustees. He was a Johnson confidant and attended the meetings of the National Aeronautics and Space Council, which were chaired by Vice-President Johnson. He attended these sessions because he was reputed to be among "those responsible members of the public" whose advice was sought on national space policy (Welsh, 1968:12).

Once NASA chose Houston, several interesting incidents immediately followed. In December, 1961, Brown and Root received the $1.5 million contract for architectural design work on MSC (Houston, February, 1962, p. 26). In 1962, Rice in turn received $192,000 from the federal government in fellowship money for graduate students in the physical sciences (Houston, September, 1962:33). Humble Oil began promoting a 15,000 acre industrial-commercial-residential development adjacent to the MSC facilities (Houston, March, 1962:46-49), the residential portion of which was expected to house 18,000 residents (Texas Magazine, August 31, 1964:77). By March, 1964, plans for a 7,200 acre industrial district on Humble's property were also well under way. The latter was reputed to be capable of including $900,000,000 in new plant investment with Lockheed Aircraft having purchased the first 500 acres (Houston, March, 1964:22).

On September 19, 1961, NASA administrator James Webb announced that Houston had been selected as the MSC site. At the time, Johnson described the MSC as "the greatest thing for Texas and Houston since the Ship Channel," and Thomas predicted that it would create a "small revolution in Houston business life" (Houston Post, September 20, 1961). It would seem that the MSC did have some impact on Houston's economy as a major economic growth spurt followed during the next decade. Between 1963 and 1967, there was a 50% increase in value added to manufactures, with a 45% increase occurring between 1967 and 1972 and a 118% increase between 1963 and 1972 (U.S. Bureau of the Census, 1971:44-21; 1976:44-21). Houston also became an immediate recipient of NASA contracts. In 1964 Houston area firms received approximately $11 million in NASA contracts (Houston, September, 1964:20-21), and by 1972 the total had increased to $262 million (U.S. Community Services Administration, 1973). Between 1967 and 1972, the number of manufacturing establishments increased from 2,621 to 3,169 (U.S. Bureau of the Census, 1976:44-21). Corporate headquarters likewise began to relocate in Houston, with 150 such moves taking place between 1967 and 1973. The moves were made ostensibly because of Houston's attractiveness as a "technological capital" (Houston, September, 1974:49). Although this growth might have occurred without the MSC, it would seem fair to say that its acquisition certainly did not hurt.

It would seem that a behind the scenes entrepreneurial booster campaign was being waged to protect Houston's interests, with Johnson, Thomas, Brown, and Davis playing the pivotal roles. However, Houston entrepreneurship in 1961 and entrepreneurship in 1910 differ on one important point. In 1910 Houston entrepreneurs had to *create* a locationally advantageous infrastructure, one that would foster economic growth. By 1961 Houston's location advantages were already existent, and it was up to contemporary Houston entrepreneurs to *exploit* these advantages. In the case of the MSC site selection process, Houston entrepreneurs could not assume that NASA's site selection officials would automatically be aware of the city's numerous advantages. Successful entrepreneurship, therefore, involved presenting Houston's case in a clear, forceful, and attractive fashion. This was crucial because other cities also sought the MSC. U.S. Representative Olin Teague, a member of the House Space Committee, indicated that Tampa, Florida, was a prime contender and that Dallas and Corpus Christi were also under serious consideration (Dallas Morning News, August 24, 1961). Entrepreneurs from these cities were also attempting to convince NASA officials that their respective communities would best serve the MSC. In Houston's case, entrepreneurship included fully exploiting the political wallop of Johnson and Thomas. When NASA officials stated that the MSC would require at least 1,000 acres for testing purposes, local entrepreneurs made certain that NASA could have its 1,000 acres. In other words, while entrepreneurship in Houston at the turn of the century meant providing the city with infrastructural advantages, in the 1960s it meant displaying those advantages so they would not go unnoticed.

Houston is an important example of the entrepreneurial impact on the Sunbelt frontier. Houston capitalists were, for the most part, barrier busters; they were entrepreneurs. Men such as Cleveland, Garrow, Peden, Brown, and Davis undertook innovative city building strategies for enhancing Houston's economy. They undertook these efforts because they felt that to do otherwise would ensure the city's economic decline. Intrinsic to their strategic success, however, were linkages with politicians like Ball, Wolters, Johnson, and Thomas. Natural conditions did not dictate the future success of cities like Houston. The entrepreneurial behavior of a few prosperous capitalists did.

NOTES

1. This is an extension of Schumpeter's (1955:74) definition of an entrepreneur as one who carries out "new combinations of the means of production." According to Schumpeter, the function of the entrepreneur is to alter the mode of production by utilizing an invention or innovative production process to produce a new product or to more efficiently produce a standard one. Schumpeter goes a bit too far when he claims that the entrepreneur is spurred not by the "fruits of success" but rather by "success itself" (1955:93). Thus, he seems to regard the pursuit of innovation as an end in itself and not as means to an end.

Marx, on the other hand, indicated that innovative acts are means to an end—the achievement of surplus value in the production of a commodity. For example, he argues that the capitalist must innovate if he expects to receive greater surplus value from his labor supply. He must discover and use ways of producing a product which will enable him to produce more while minimizing labor costs. Once this is done, "the value of labour-power can be made to sink, and the portion of the working day necessary for the reproduction of that value shortened" (Marx, 1967:315).

The above interpretations, however, confine entrepreneurship to the factory floor or the inventor's workship. According to my interpretation, entrepreneurship involves creatively using any societal resource—including government—to turn a profit. It is thus not purely economic in its application.

2. For the entrepreneur the parameters of accumulation possibilities are not nearly as well defined as they are for the nonentrepreneur. What an ordinary capitalist perceives as an impenetrable barrier to accumulation, the entrepreneur views as an inconvenience whose conquest will yield greater accumulation possibilities. Hence, the barrier notion is a phenomenological concept which varies according to the "eyes of the beholder."

3. Albion (1939:91) argues that hinterland trade would probably have come to New York anyway. In his view, the Erie Canal certainly stimulated the city's commerce, but its chief importance lay in fostering commercial settlements in the West. This argument contrasts with Rubin's (1961:8), who indicates that the canal allowed New York City to monopolize western trade to the disadvantage of other Atlantic Coast cities.

4. Houston members were: Ball, Cleveland, John Browne (owner, Browne Commission Company); J.F. Meyer president, Joseph F. Meyer and Company); Matthew Nicholson (chief clerk superintendent, Texas and New Orleans Railroad); C.H. Milby (proprietor, Harrisburg Brick Company); H.F. MacGregor (real estate operator); C.L. Bradley (attorney); E.A. Peden (general manager and vice president, Peden Iron and Steel Company) (Houston Post, June 10, 1910; Houston City Directory, 1910-1911).

The executive committee was also responsible for establishing electoral procedures. The voters had to approve both the organization of the Harris County Navigation District and the $1,250,000 bond issue. The committee decided that these questions could not be divorced and worded the ballots in such a way that a favorable vote signified approval for both the organization of the navigation to include all Harris County *and* the $1,250,000 bond issue. This will be referred to as either the Ship Channel or Navigation District issue.

5. The occupational backgrounds of the leaders of the North End Improvement Association indicate its working- and middle-class character. George Krengel, the organization's president, was head janitor for the Houston Land and Trust Company. Backgrounds of other Fifth Ward activists were similar.

Also, Houston's Fifth Ward—the city's working-class district—was separated from the central business district by the White Oak and Buffalo Bayous. Although bridges crossed these two streams, they provided only indirect access to the downtown, and members of the association had tried for five years to acquire a more direct route. The Main Street Viaduct, which would traverse the bayous at their confluence, was sought for this purpose. Its estimated cost was $400,000 (Houston Post, August 5, 1910).

6. U.S. Representatives Olin Teague (Bryan-College Station, Texas) and Robert Casey (Houston) were both members of the House Space Committee. The MSC was built in Casey's district, but Thomas (who was also an alumnus of Rice University) was more influential in the selection process.

REFERENCES

ALBION, R.G. (1939). The rise of New York Port. New York: Scribner.

BALL, T. (1936). The port of Houston: How it came to pass. Houston: Houston Chronicle and Houston Post.

BOEHM, R. (1975). Exporting cotton in Texas. Austin: University of Texas Bureau of Business Research.

DUNCAN, B., and LIEBERSON, S. (1970). Metropolis and region in transition. Beverly Hills, Calif.: Sage.

DUNCAN, O., SCOTT, W., LIEBERSON, S., DUNCAN, B., and WINNSBOROUGH, H. (1960). Metropolis and region. Baltimore: Johns Hopkins University.

FUCHS, V. (1962). Changes in the location of manufacturing since 1929. New Haven, Conn.: Yale University Press.

GOODRICH, C., and SEGAL H. (1953). "Baltimore's aid to the railroads." Journal of Economic History, 13(winter):2-35.

HEBERLE, R. (1954). "The mainsprings of southern urbanization." In R. Vance and N. Demerath (eds.), The urban south. Chapel Hill: University of North Carolina Press.

Houston (1962). (September):30-37.

ISARD, W. (1956). Location and space economy. New York: John Wiley.

JOSEPHSON, M. (1934). The robber barons. New York: Harcourt, Brace and World.

LIVINGOOD, J. (1947). The Philadelphia Baltimore trade rivalry. Harrisburg: Pennsylvania Historical and Museum Commission.

LUXEMBURG, R. (1951). The accumulation of capital. New York: Monthly Review.

MARX, K. (1967). Capital, vol. I. New York: International Publishers.

OATES, S. (1964). "NASA's manned spacecraft center at Houston, Texas." Southwestern Historical Quarterly, 67(January):350-375.

RISTER, C.C. (1949). Oil! Titan of the southwest. Norman, Okla.: University of Oklahoma Press.

RUBIN, J. (1961). Canal or railroad? Philadelphia: American Philosophical Society.

SCHUMPETER, J. (1955). The theory of economic development. Cambridge, Mass.: Harvard University Press.

TURNER, F.J. (1956). "The significance of the frontier in American history." In G.R. Taylor (ed.), The Turner thesis concerning the significance of the frontier in American history. New York: Holt.

U.S. Army Corps of Engineers (1901). Annual report upon the improvement of certain rivers and harbors in Texas, Appendix U. Washington, D.C.: U.S. Government Printing Office.

--- (1919). Houston ship channel, Texas. U.S. House of Representatives, 65th Congress, 3rd Session, Document 1637. Washington, D.C.: U.S. Government Printing Office.

U.S. Bureau of the Census (1971). Census of Manufactures, 1967, Volume III, Area Statistics, Part 2. Washington, D.C.: U.S. Government Printing Office.

--- (1976). Census of Manufactures, 1972, Volume III, Area Statistics, Part 2. Washington, D.C.: U.S. Government Printing Office.

U.S. Bureau of Statistics (1884). Report in regard to the proposed improvement of the harbor of Galveston, Texas. U.S. Senate, Misc. Document No. 111, 48th Congress, 1st Session. Washington, D.C.: U.S. Government Printing Office.

U.S. Community Services Administration (1973). Federal Outlays, City Summaries, 1972. Washington, D.C.: U.S. Government Printing Office.

U.S. National Aeronautics and Space Administration (1969). NASA Administrative History. Unpublished manuscript.

WARNER, S.B. (1968). The private city. Philadelphia: University of Pennsylvania Press.

WELSH, E. (1968). Interview for the Lyndon Baines Johnson Oral History Project. T.H. Baker, interviewer. Austin, Texas: unpublished transcript.

5

The Selling of the Sunbelt: Civic Boosterism in the Media

GENE BURD

□ FROM MANIFEST DESTINY TO MARLBORO COUNTRY, the journalistic image of American urban and regional expansion has glowed with the optimism of the superlative. Booster newspaper language was important in the building of the frontier towns (Dagenais, 1967), and the rise of modern photo and electronic journalism expanded vision and values in urban symbolism (Burd, 1977).

As crucial agents of urban growth and development (Burd, 1969a, 1969c), the communications media define and delimit urban areas, develop urban problems for the civic agenda, and dutifully defend the reputations of urban empires as desirable sites for economic and residential relocation and expansion (Burd, 1970). The hallmark of media content has been peerless boosterism: congratulate growth rather than calculate its consequences; compliment development rather than criticize its impact (Devereux, 1976; Burd, 1972, 1969b; Freidel, 1963).

The presentation, promotion, and purchase of the Sunbelt cities continues to reflect the role of media as both chronicler and cheerleader in urban society. Escapees from older Northeastern cities read travel pages which promote nomadic wanderlust in Sun Country and tourist agencies peddle a fantasy world where the sky is blue, the sand white, and the palm trees green (Sesser, 1970). Real estate sections sell urban promises rather than tell of urban problems (Kuhn, 1966); and Henry Ford's pledge to solve big city problems by providing an auto to leave them is aided by the longtime press tie to the "car culture" (Flink, 1975). Once Sunbelt escapees arrive by auto, the press accepts an antiurban, decentralized, auto-dependent life-style (Foster, 1975), and newspapers give more coverage to sports than to urban development (Ernst, 1972).

The migration and pilgrimage to the Sunbelt was historically coded in the journalistic terminology of national progress hawked in P.T. Barnum style as the patron saint of boosters. The parade of the longest and largest, the biggest and best, the grandest and greatest stretched from the cities of the New World in New England and New York beyond the Hudson and Mississippi to the Great Plains toward the West—new, but wild and woolly and "won" despite a huge ocean, mountains, and desert; and beyond to the South—old, but Deep and Solid, and "rising again and again" as the New and Improved despite Dixie's defeat from the Border States to the Gulf.

Newspaper editors were salesmen for the South and West long before irrigation and air-conditioning made the deserts and swamps palatable, and long before television, country-western singers, and advertising agencies made the Sunbelt popular.

Horace Greeley, editor and founder of the New York *Tribune,* popularized the advice, "Go West, young man, go West" in the mid-19th century, and his own agriculture editor founded Greeley, Colorado. A similar booster, Henry Grady, editor of the Atlanta *Constitution,* publicized the term "The New South" in his 1886 speech in New York City (Grady, 1966), where he assured businessmen that slavery and secession were dead and the South could attract industry and new urban growth.

That was soon after the press hailed the nation's Centennial, and asked for national unity and rediscovery of the South, which had suffered from a bad press before (Green, 1972) and after the Civil War and a negative journalistic bias except for some metropolitan journalists from the East who visualized Southern political and industrial expansion (Floan, 1958). By the mid-20th century, the expected capital and immigration to the South had not fully materialized and the adverse image was cultivated by critical journalists like H.L. Mencken who ridiculed the South's "shoddy cities" in the 1920s "as sterile, artistically, intellectually, culturally as the Sahara Desert" (Egerton, 1974:207). A Mencken article in the *American Mercury* magazine in 1931 on "The Worst American State" ranked the South the lowest and referred to the Cotton Belt as "the least advanced" part of the U.S. and inhabited with "hordes of barbaric peasants." (Mencken's subjective paraphrase of the more objective Cotton, Corn and Wheat Belt terms created the "Bible Belt," a derisive reference to the Southern power of Protestant fundamentalist churches and preachers.)

In the 1930s, while the public image of the West moved from the portrayals in the Lewis and Clark Journals toward the "Grapes of Wrath" and Hollywood Westerns, the South's liveability was again doubted in 1938 when the Roosevelt Administration enraged Southern editors by designating it as the nation's number one economic problem for the National Emergency Council. Northern news focus on Southern school segregation in the 1950s and the Civil Rights Movement in the 1960s only added damage to the South's image, tourism, industrial development, and pride. *Time* magazine was called "one of theprime South-baiters" which helped to prevent the South from getting a "fair hearing in the market place of national public opinion" (Workman, 1960:62).

The Space Age perhaps widened the vision of journalists as time and space were shrunk by transcontinental jets, the TV nets, and satellite weather maps which expanded the communicated images of community. Township, ward, borough, village, town, city, county, and state (up, down, and out) became less meaningful than central city, suburb, metropolis, and region, which spread further into strips, crescents, rims, belts, and other configurations shaping the new urban areas (Clay, 1973).

Transportation was shifting people and communication was shifting their pictures of desirable places to live. In 1961, San Diego journalist Neil Morgan wrote of 11 Western states as a "new sub-nation" which would eventually dominate the country politically as people moved from the East to escape urban crowding for the "scenic extravagance of the rawboned West" (Morgan, 1961). In 1963, West and South may have moved closer in time and personality as media focus on the Southwest and the Great Society merged on President Johnson and Texas, where geographical and sociological boundaries of the South and West appear to meet. There, cattle and oil replace cotton, hillbilly and country Western blend, religion and racism compromise in the aftermath of Mexican colonialism and the Confederacy, and fauna and flora change as plantation and desert frontier consolidate the hopeful South and the unfinished but optimistic West (Vandiver, 1975:47).

By the time the 1960s ended, journalist and political analyst Kevin Phillips had coined and christened terms like Sun Country, Sun Culture, and Sunbelt to denote the alloy of a new majority of political conservatives who grew out of the new social affluence, leisure, and mobility, and who had found a political and geographical home in the South and West (Phillips, 1969:437). The Sunbelt thesis was reiterated a few years later by a more critical journalist, Kirkpatrick Sale, who called the new concentrations a "Southern Rim" and a regional "Power Shift" challenging the Liberal Eastern Establishment from an economic base of technology, defense, oil and gas, aerospace, agribusiness, and electronics, as well as the warmer climate (Sale, 1975).

The New and Improved South became a walking advertisement packaged with the good will of Jimmy Carter, the good ole boy style of his brother Billy, and the good life of the small town (Plains, Georgia). It created good news copy (Van Hetherly, 1977) as it sold to urban Americans (Martin, 1977) the mythical virtue and agrarian simplicity of rural, crackerbarrel wisdom.

"A New South and a New Coalition Elected Carter," proclaimed a November 7, 1976, Milwaukee *Journal* story by Alabaman Leon Hughes, with the President-elect grinning in front of the Atlanta skyline. "We Ain't Trash No More" wrote Southern journalist Larry King in the November *Esquire.* Southerner Roy Reed welcomed Southern revenge (New York *Times* magazine, December 5, 1976) against Northern visitors who once "marked their supremacy by spitting scorn. And the scorned are still spitting back. We the scorned have always felt a special need to put the best face on things. At its most primitive level, that is the reason for the Southern grin." (And what's better than a smile to usher in the Sunbelt?)

GRADY AND GREELEY REVISITED: THE SUNBELT BLISS BLITZ

If editors Henry Grady and Horace Greeley were alive for the Carter victory and the Bicentennial, Grady would find an even Newer South which would have him "dancing editorial jigs" (McGill, 1963:208), and he would find the Atlanta Chamber of Commerce opening a New York City branch office to conduct a $1 million advertising and public relations campaign to lure Northeastern industry to his "city without limits" (King, 1977).

Editor Greeley would find his namesake town of Greeley nudged by Denver, whose *Post* on page one daily proclaims the city "The Climate Capital of the World" and the paper "The Voice of the Rocky Mountain Empire," where singer John Denver's musical Rocky Mountain hype has sold more of the West than his own editorials and more publicity for Colorado "than Cleopatra did for Egypt, the Eiffel Tower for Paris or the Taj Mahal for India" (Golab, 1976:38).

Both 19th century editors would find their genteel speeches and editorials for Southern and Westward expansion replaced by the "hard sell" and huckstering of "Sunbelt salesmen" armed with "the brash, new money of the South," "tall and proud," "politely aggressive," "But unlike the swashbucklers of old, their pirate ships are subtle and sophisticated advertising campaigns and promotional trips. . . . Their weapons are clean air, quality of life, air transportation, largely non-union work force and the spirit of economic regeneration" (Tempest, 1976).

Greeley's *Tribune* and other Manhattan neighbors are long gone, but the *Wall Street Journal* in 1973 carried ads urging Northern industrialists to "Leave your problems behind, try Austin, Texas. The weather's mild, the atmosphere is great for year-round tennis, for higher profit/fewer problem business operations." "Were Horace Greeley around today, his advice would be largely unchanged except for the direction he turned the nation's young men toward. Fortune lies today to a great extent in the South, the nation's 'Sunbelt', symbolized not by grits, but by growth" (Frink, 1977:G1). So goes the editorial legacy and litany of growth.

Fifty years after the November 1925 *Literary Digest* ranked only Los Angeles and New Orleans outside the Northeast as desirable places to live, H.L. Mencken and others would find in the January 1975 *Harpers* article on "The Worst American City" that only one Northeast city ranked in the top 13 of the 50 largest U.S. cities desirable for health, housing, affluence, education, and other quality of life amenities. The journalistic value shift to the Sunbelt was evidenced in the August 1975 *Ladies Home Journal* rank of 15 "America's Best Suburbs," 10 of which were in the Sunbelt. *Family Circle* of February 1977 unloaded its belt-booster by devoting a whole issue to the Sunbelt including a ranking of "22 Places in The Sunbelt Where There's Good Living."

Prosperous regional booster magazines (*Sunset, Southern Living, Texas Monthly, Arizona Highways*) extol Sunbelt virtues and life-styles (Allen, 1974, Burd, 1974; Brown, 1972); and powerful regional Sunbelt dailies like the Los Angeles *Times* and Houston *Chronicle* extend their tradition of civic activism to aid the power structure in economic development and urban rivalry (Pollak, 1967; Gordon, 1965; Gibson and Greene, 1973; Flippo, 1974). Such regional print media and the increased power of television to cross old geographical and political boundaries makes regions somewhat less dependent on national media for favorable images and promotional goodwill. However, the Gradys, Greeleys, and Menckens of the East retain power in the form of the New York newspapers and newsmagazines. Their national agenda-setting prestige and distinction are still sought by regional boosters, and by the mid-1970s, it became evident that the Northeast press was shifting from a negative to a positive news image of the South, and to the eventual acceptance of the Sunbelt as a possible major regional entity.

For example, from 1970-1975, in the *Wall Street Journal,* the South got a bad press with 15 articles each year (1970-1972) dealing mainly with racial problems (school segregation, Black voting rights, and integration) and sporadic stories on cotton, tropical storms, and Disney World. In the next three year period, the *Journal* indexed only 16 stories on the South (seven in 1973; five in 1974; and four in 1975) covering more race relations, moonshining, guns, bourbon, catfish, Mississippi mud, peanut subsidies, and Southern efforts to get Southern media to improve reporting on Southern problems.

By early 1976, it was clear that the growth of the South would bring improved publicity in the Northern press. The day after the updated U.S. census figures indicated that the South and West were shifting the population center of the country, the New York *Times* on February 8-13, 1976, in a series of six page-one articles[1] by six writers in six days legitimized the historic shift to the Sunbelt. By its national agenda-setting power and its esteem as a national newspaper, the series precipitated a "press shift" in the form of reprints and further 1976 articles in national publications reaching mass audiences far beyond the works of Phillips and Sale, who had helped coin the Sunbelt idiom.

The tone and theme of the *Times* articles was predominantly boosterish rather than critical, with negatives submerged and sublimated among superlatives; and with the Deep South, rather than the West, presented as *the* Sunbelt. That emphasis (in the Carter election year) dominated the news angle of 1976 national media coverage of the Sunbelt.

The writers recited the ritual of growth advocates, with subdued warnings of adverse consequences of growth religiously sandwiched among or at the end of recitations of the bonanza of boom towns.

The writers told of the leisurely, mobile, affluent lifestyle, the expanding job and labor pools, the lower wage scales and low taxes, the freedom from blight in the depressed North, the escape to the sun away from high fuel costs and Northern social tension, "the decline in the importance of the race issue in the South," the open life, open space, and open collars of Yankees learning drawls and drinking Coke with bourbon; the warmth and trust among Southerners, the freedom from the Northeastern rat race and congestion, the relaxed life of back slaps, year-round vacation spots, the disappearance of the carpetbaggers and "meddling integrationists," the casual manners and elbow room, the weak labor unions, the low cost of living, low property taxes, and, of course, air-conditioning, without which, as the Houston mayor said, that city "would not have been built—it just wouldn't exist."

As with subsequent national media stories on the Sunbelt, the negative aspects were often veiled and muffled: a subdued reference to the "hardships of growth" like traffic congestion and strip housing; a cloaked paragraph in a glowing booster story on Houston, with a cryptic phrase by a labor official on how Texas is number one in climate but number one in poverty (32nd in state per capita income); and how "Blacks, Mexican-Americans and many poor rural whites remain, at least, marginal participants in the regions' boom. Their cheap labor, in fact, is one of the Sunbelt's attractions," said the *Times* story headed: "Houston, as Energy Capital, Sets Pace in Sunbelt Boom."

Nonbooster references were screened by glowing heads, and final paragraphs on "the need to discipline and coordinate the growth that is already creating pollution and other social environmental intrusions upon the quality of life the new migrants seek"; plus shrouded admissions that damage from lack of growth controls, were "doomed to be repeated" in the area "before a meaningful groundswell of public support can be mastered" to control growth. One last paragraph of a story noted that the Sunbelt movement was "the last migration into relatively unused open space," which had traditionally been overrun, battered, and then abandoned, but ended with a hopeful presumption that "the next cycle of mass movement will be one of reclamation: to rehabilitate the great metropolitan centers that were allowed to succumb to the same social malaise now being hailed as the inescapable by-product of progress in the Sunbelt."

The 1976 national Bicentennial and the Carter campaign was a time for media indulgence in superlatives in a search for national self, regional margins, and for national leadership after Watergate sins and perhaps the long guilt over the Civil War drawing press attention to a region long wrestling with its own identity. In addition to the *Times* series, the *Wall Street Journal* in 1976 flowered with 13 items on the Sunbelt, proclaiming its rising economy, its new image, the shift of power, and how the Carter election "emancipated the region from the political bonds of the past." The Christian Science *Monitor* blazoned the bliss of the area with a 60 inch, two page spread (Moorhead, 1976), showing glistening skycrapers and skylines, the "impressive growth" and "muscular progress" of Southern cities, and Birmingham's "fresh face is a sharp contrast to its enduring reputation as a dirty, socially troubled industrial city." The *Monitor* propagated the glad tidings that the region had "a work force with a reputation for diligence, and favorable tax treatment have helped spur the industrial growth," but a fragment of one sentence in the 60 inches of copy bared that in New Orleans, "more people live here at or below poverty level than just about any other U.S. city." But the comfort of the South was pronounced on luxuries such as air-conditioning which "has taken the sting out of climate and enticed many Northern individuals and companies" (Moorhead, 1976:14-15).

After the Carter nomination, the terms South and Sunbelt became interchangeable in some of the national magazines, and after his election, which the West largely opposed, Carter Country momentarily became Sun Country for some journalists careless about regional definitions, but enthusiastic for new rhetorical devices.

The overlapping themes of the New South and the Sunbelt were evident in two September 1976 issues of *Saturday Review* ("The South as the New America," September 4, 1976), and *Time* ("The South Today: Carter Country and Beyond," September 27, 1976). *Time* used 70 journalists, many with ties to the South, to write and report about 11 states (Virginia, the Carolinas, Tennessee, Georgia, Alabama, Mississippi, Louisiana, Arkansas, Florida, and Texas). The issue was a chamber of commerce dream: overwhelmingly positive, upbeat, lavish in booster praise, while any negative aspects were sparse, buried, appended, or carefully qualified with an apologetic, euphemistic tone.

Time, hardly the accused "South-baiter" anymore, told of "The Good Life" in six color pages on recreation and play, and of "Those Good Ole Boys" with "a frame of mind based on the premise that life is nothing to get serious about" and "What he really loves is his automobile," more than his woman. The magazine also colored a flattering picture of the "new class" of Southern politicians "out of a cocoon" and purified by the increase in Black voter registration and votes. (*Time* hinted at cautious doubt of Carter: "How Southern Is He?")

On issues of race and violence, *Time* softened the negative. It said that paradoxically in the South, "violence is not far from the surface," then added "But up North, the combined rate of violent crimes (murder, rape, aggravated assault, and robbery) is still greater than that of the South." It pictured the South as moving "Away From Hate" with "greater strides than the rest of the country" and that in Birmingham, "A City Re-Born," "Bull Connor has since died—and so has Birmingham's bitterness." Although the admission is made that desegregation is "An Unfinished Task," one expert is quoted as saying that its record "is better than the North's," and *Time* reasoned that the South "has borne the brunt of federal court orders" and "Throughout the South, news of Northern and border-state unrest over busing has been greeted with understanding and something more than a little regional hubris." (A black *Time* correspondent in Atlanta wrote of "Segregation Remembered" and conceded that "The Southern white man, even at his most bigoted, always had some noble impulses: loyalty, independence, courage.")

The Northern-based magazine adopts a stance of forgiveness and absolution for the South's shortcomings, as it excuses and exempts that which does not boost the new image. In telling of "Small Town Soul," "Times are rough right now . . . puny drought-burned tobacco leaves," few store customers (Chatham, Va.) and "many farmers are turning to moonshining whiskey, but "EVEN SO, there is a basic optimism." The economy is "Surging to Prosperity" with "Those Brash New Tycoons" and "The Non-Stop Texas Gusher," but "Progress always exacts a price . . . [and] prices in the South last year rose . . . THOUGH they are still at a lower level than in the

North." In a brief segment on unions, "You Gonna Get Fired," "frustration is the rule for Southern unionists . . . [but there is] . . . fierce employer resistance, BACKED OFTEN by local public opinion." (Capitalization added for emphasis.)

Time's acquittal of the South ends with an essay by Yale's Southern historian C. Van Woodward, pleased that his native region is losing its inferiority complex and may reclaim from the North the power to run the U.S., which the South did for 32 of the country's first 36 years. He said the "old grounds for Northern moral superiority" gave way with Boston racial disturbances, which "helped to modify the old stereotypes and mitigate fear of Southernization in the North." He said "a new type of Southern patriot took delight in pointing a finger at Boston, that oldest moral critic and accuser of the South." In what might seem an indirect commentary on the public before the election and on the Northern post-Watergate press giving the South a better press, Woodward commented:

> Self-righteousness withered along the Massachusetts-Michigan axis. Northern morale was further lowered by Viet Nam and Watergate, devastating blows at the widely held myths of invincibility, success and innocence. Those myths were never shared by the South anyway. In their present state of disenchantment and demoralization, Northerners are apprehensively looking south for leadership.

As for some fears that the North will bring urban homogenization and destroy Southern values, Woodward argues that "only the most vulgar economic determinist would argue so." He brushes this off saying that "Sunbelt opulence still leaves the South much the poorest of the country's regions. The old Southern distinction of being a people of poverty among a people of plenty lingers on. There is little prospect of closing the gap overnight." That problem was muffled and masqueraded in the parade of pleasure and felicity in the *Time* issue.

An unparalleled usage of transcendent superlatives inundated the special September 4, 1976 *Saturday Review* issue on "The South as the New America," as Atlanta, Houston, and New Orleans were presented as "Supercities" symbolic and representative of the Sunbelt and dominating that region as well as the magazine edition. The adverse or negative consequences of urban Sunbelt growth are either ignored or disguised and distorted.

As with the *Time* edition, the press enthusiasm for the Carter story warped the Sunbelt definition to fit the South, and the images once again glorified escapism while exalting urban growth. The magazine acclaimed the "supercities exploding into Southern skies" and "These once-drowsy towns take on the look of a stage set for 'Star Trek' . . . like Tomorrowland spawned today" as non-Southerners seek a "more comfortable place to live" and industrialists find out about "January golf," and tourists flee urban centers to find that "Plains is Abilene, Kansas, or West Branch, Iowa" (presidential home sites).

In one article, the Houston *Post's* columnist Lynn Ashby said his city had "boomed and continued to boom" as "a better-mousetrap sort of town" and was "the golden buckle of the Sunbelt," in a borrowed phrase from the *Wall Street Journal.* "Houston," said Ashby in worshipful esteem, is "a catchword for a modern growing metropolis where anything is possible and anyone can try anything—at least once."

Ashby not only camouflaged typical Houston problems like pollution, crime, and congestion, but led readers toward the escape hatch of space flight. "Houston, the first word from the moon," he wrote, "is the last word in American cities. Big, strong, young, insufferably confident, Houston is rushing hell-bent into tomorrow without much thought about the day after." Ashby admits that Houston has neither "arrived" nor is "finished," but "the trip is exhilarating" and in his praise of the Lyndon Johnson Space Center he exalted, "day to day it is a little humbling to know that your neighbor has putted a golf ball on the moon."

He praises the Astrodome (which has severe financial problems); the lack of city or state income taxes, and low property taxes (in a state with few social services and many social problems); and the "friendliness of the natives" (in a city which has a notorious crime rate). He calls Houston "aggressive, unpretentious, adventuresome, given to Saturday night specials and Sunday morning religion . . . an open, loose city, where all men are created equal if they can afford it. . . . You, too, can make a bundle if you are willing to pay the price."

In a flippant and macho manner on crime, Ashby wrote that "until the new penal code went into effect three years ago, the law allowed a man to kill his wife's lover but not his wife, and at least one husband went to jail for shooting at the lover but killing his wife instead. In Texas, poor marksmanship can be a punishable offense." In a similar cocky vein, he boasted that the "downtown area, far from caving in to urban blight, is being built up, and is growing in every direction" and "the 207 miles of freeways make it quick and simple to get around town if you avoid the deadly rush hour traffic jams."

The same vainglorious attitude pervaded an article in the *Review* by David Chandler on New Orleans, which was described mostly via commendation of the Superdome, which was pictured as "a physical and metaphoric symbol" of New Orleans and as the "largest finished building in the world" into which could fit Notre Dame, the Pantheon, St. Paul's (or the Astrodome), and even St. Peter's–if the Vatican tower were not so tall. He admitted that there had been "hints of corruption and fraud" which had "clung" to the dome but that "nevertheless, it survives and now it works," and when one air passenger saw it from the sky, the "city looked on the verge of lifting off, of making departure for, say, Venus."

The selling of the Sunbelt city of New Orleans includes an account of the new hotels and convention centers and earthly escape from urban problems through urban frivolity. Chandler concedes that New Orleans "has been mired in corruption, fraud and poor management throughout much of its history," that its streets are "the dirtiest in America," and that its "drinking laws are virtually nonexistent"; BUT he says that tourists and conventioneers are attracted to "The Big Easy"–New Orleans–"the most wickedly charming of American cities." "If you are in step with her vibrations," wrote Chandler, "she is soft-faced and loving, full of delightful surprises, and very, very permissive."

In 1976, during what appeared to be summit year for selling the Sunbelt, *U.S. News & World Report* published numerous stories which were, in effect, "advertisements" for the Sunbelt: "America's 'Sunbelt' Is Growing in People and Power" (April 12, 1976); and "The New South: Pushing Forward on All Fronts" (August 2, 1976). Seventy-eight inches of the 90-column-inch April story told about the "boom," "rapid expansion," the "major shift" in business and jobs, the lower living expenses, the new wealth and political influence of the Sunbelt; "air-conditioning has made hot summers bearable," "a more relaxed life style that contrasts with the crime and congestion of many Northern Midwestern cities"; and "right to work laws" as "another attraction for management." The emphasis is on regional pride as "the South and West, after years of lagging behind, are ready to move out in front."

At the end of the long story, 12 inches were devoted to how Sunbelt residents view such growth. Although there was an admission that "they worry increasingly, however, about the need to avoid making the same mistakes that plague many Northern communities—overloaded tax systems, unplanned urban sprawl, excessive pollution and other hazards," the magazine emphasized that "proponents of continued growth are prevailing," "growth is accepted as inevitable," many officials of the South and West "fear that the region may not be able to throttle back its drive for expansion, no matter what harm ultimately result." *U.S. News* concludes by wrapping the growth ethic in civic pride:

> It's a welcome turn for most people after years of living in what others saw as an economic, political and cultural backwater. . . . All the headaches, however, cannot subdue the elation and pride that most residents of the Southern and Western U.S. feel when they look at the growth seen in years ahead. For them, after decades in the rear ranks, it's nice to be out in front.

In the August 2, 1976, article, the concession is made that such economically based growth is "a direct result" of the civil rights movement which many Southerners fought; and "business leaders say it would have been impossible to persuade national corporations to move into the racially repressive, often violent atmosphere that pervaded the old South." The magazine reports that times have changed as racists are a few "diehards," Baptists are involved in social problem-solving, schools are "the most fully integrated in the country"; and the "white-supremacist groups are dying anachronisms," as a small story at the end reports: "KKK: A Fading Remnant of the Past."

The South, *U.S. News* reported, today looks like any other region: "The same freeways, fast-food franchises, office towers, shopping malls, and suburbs . . . like all big cities, are plagued with smog, traffic congestion, noisy airports, crime and 'porno' shops." As for the lament that the South's unique, tranquil, folk society will be homogenized by urbanization, *U.S. News* excused the adversity of change, saying that:

> Most Southerners, however, seem willing to endure a few traffic jams in exchange for steady jobs and neat homes in the suburbs. Air-conditioning has made the increasingly hectic pace of Southern life more bearable. [p. 45]

What's more, the magazine quoted the Houston mayor saying that the South would lose "very little that was good," and Mississippi leaders as "tired of being last in everything."

The mass media have defined the Sunbelt with the cliches of well-being and being Number One in the sunshine of good fortune rather than accented the clouds of regional adversity. The Belt's bases and criteria are frequently not explicit as geography, meteorology, geology, and sociology are mixed indiscriminately by journalists in a blast of bombast, saccharine sales pitches, and simplistic and optimistic epigrams, which are constant, repetitive, alliterative, and combine the poetic civic ideology of the chamber of commerce with the self-fulfilling prophecy of the press.

MEDIA MERCHANTS MARKET SUN COUNTRY

The Sunbelt bathes in media power puffery where image and reality blend in the promotion and purchase of the Promised Land. Journalists spread the zest and gusto, the rapture and delight, the cheer and joy, and the cloudless complacency of the "pleasure belt" ripe for safe escapism from the older cities and ecstatic release from the Northeastern urban density.

Indeed, in Sunbelt journalism, it is difficult to separate ads and news. So-called "objective" stories mix the metaphor of promotion with facts. The "straight" news story on "Cities Where Taxes Are Highest–and Lowest" (*U.S. News,* June 6, 1977, p. 66) becomes a Sunbelt advertisement for luring business and people from the North, where taxes and living costs are higher. Likewise, the simple news feature story "sells" the Sunbelt, as when the small town of Hondo, Texas, got inquiries from hundreds of prospective new residents, when its mayor asked President Carter for imported unemployed loafers so that the town would have enough unemployed to qualify for a federal grant to build a new civic center (*Newsweek,* June 27, 1977, p. 21).

The media collage of the Sunbelt mixes the mythology of siesta and the plantation with the newer romance of conquering the desert and harnessing the sun. Hollywood plastic juxtaposes Manhattan and "McCloud" and permits Southerners to recall rural hardship on "The Waltons," as Optimism Unlimited reigns over the new sunny Garden of Eden, where the Good Life makes good copy in the New Nirvana.

The Sunbelt Utopian vision appeals because it means castles replace swamps and snakes; and extravaganzas make the desert bloom, to "Win the West" and find the Pot of Gold at the End of the Rainbow. The Sunbelt is the Gold Rush, "California or Bust," "Eureka" ("I've Found It!"), Brigham Young ("This Is The Place"), and "GT" ("Gone to Texas") all rolled into one frontier rhapsody. It is the essence of the news story—the search for the New Jerusalem, the City of God, the Fountain of Youth; the reclamation of the American Sahara, the rise of Atlanta and Phoenix from ashes, the ascent of Alburquerque from atomic blasts, the Manifest Destiny of Arizona, the "Boomer Sooners" of Oklahoma, the settlement of the suburbs, exurbs, and the moon; and the domination of the uncivilized, undeveloped, undiscovered, unknown with unlimited freedom. It is masculine man conquering nature, as one Sunbelt news story extends the glorification of those who exploit natural resources:

> The swaggering Texas oilman is a character as familiar to American folklore as his cream Stetson and two-toned inlayed pointy-toed boots. . . . But how about the swaggering Texas coalman? Or the swaggering geothermal scientist? [Cryer, 1976:B1]

Success in the Sunbelt means the rags to riches tale American journalists have told for years from the Dust Bowl to the Rose Bowl, from the chance to gamble from derricks in West Texas to tables in Las Vegas. It is the fast buck, the faster horse, and even faster cars to speed the getaway provided by Henry Ford and the escape to the Prairie House and ranch lawns of Frank Lloyd Wright and the breezes of the Gulf and Pacific just like the travel pages promised. The press promotion of suburban escape and the idolatry of small towns extends to the selling of the antiurban rural myths of the West and New South. It is the flight from Depression and Recession to the Land of Milk and Money, from Midwest snow and Great Lake winds to "Little Iowa" (Long Beach) and the "New Bronx" (Miami). It is the Los Angeles *Times* (March 10, 1977) featuring "A Place in Sun For Refugees From Winter" in the East, where Buffalo is snowed under, but downtown Los Angeles "sparkles under a sky swept clear of smog" (December 3, 1976). It is the Dallas *Morning News* with a page one welcome to "Refugees Find Oasis in Texas" escaping the Northern California drought (July 31, 1977). It is indeed "Sunbelt Magic" for a Houston bank vice-president writing in the Summer 1977 *Flying Colors,* Braniff's in-flight travel magazine.

The Sunbelt sells the sorcery of space—from the Mojave to Houston—and power—from the Solar Center to Boulder Dam. It is a television commercial and a Sunday supplement of a new life-style built around the realization of advertising copy, aided by irrigation and refrigeration and the easy leisure of retirement and consumption in the post-industrial society. It is the country "Los Angelized" with the suburbanized, decentralized, anticity model distributed in real estate pages, travel brochures, fashion plates, and Hollywood—with its sunglasses, suntans and sunbathing—competing with Miami sun and its own "Sunshine State."

The Sunbelt becomes the image of purity in the New Bible Belt now "Born Again," merged with the purity of orange juice from Malibu to Miami, and the sun for asthmatics, tuberculars, and sinus sufferers in Arizona and New Mexico. And the new God of the Sun *must* shine, even in drought! Sunshine is *good* news as TV weathermen equate rain with *bad* weather as they package and sell the sun to tourists and new industry. (In El Paso, the *Herald-Post* each day on page one tells readers, i.e., on June 23, 1977: "The Sun shone today for the 219th consecutive day. The Sun has failed to shine only 25 of the last 5,620 days." One El Paso radio station has even given away free advertising time on days when the Sun did not shine. In the Panhandle, the mayor of Guymon, Oklahoma, criticized TV weathermen and reporters for picturing his area with adverse weather "like Siberia," when actually, he pointed out, "we have sunshine 345 days a year and that compares with Miami, Florida," which Jackie Gleason's TV show proclaimed "America's fun and sun capital.")

The media helped merge South and West with new visions and sounds and new voices and smells. The Sunbelt is the blend of Southern pine and Western sage, the dialects of LBJ and Jimmy Carter; the macho of Billy Jack and Buford Pusser "Walking Tall"; the canned urban laughter at the rural Beverly Hillbillies and Hee Haw; the city acceptance of the Good Ole Boy and the Grand Ole Opry; the amnesty for "Birth of a Nation" and "Tobacco Road" and the new amnesia for "Deliverance," "Gunsmoke," and "The Longest Yard," as media images become Sunbelt reality in Southern movie lots of extending Hollywood to Texas and Georgia (Henrickson, 1975).

The Southern worlds of Mitchell, Caldwell, Williams, Faulkner, and Capote are said to be "Gone With the Wind," and yesterday's Thomas Wolfe, homesick and haunted by trains, realizing "You Can't Go Home Again" to North Carolina, is replaced by today's Tom Wolfe, chic and clever urban observer whose jet-set "home" ranges from Las Vegas to New Orleans to South Carolina. "Everybody wants to be a Southerner now. . . . They want our mores, our truths, our fried chicken," Southern poet James Dickey told a reporter at the Carter inauguration eve, where he read his poem, "The Strength of Fields" and mixed with John Wayne, symbol of the strong man on the Western plains (Axthelm, 1977:25).

The media merchandising of Marlboro Country and the parade of Miss Americas, predominantly from the Sunbelt, advertises the nostalgic and rural frontier life of the open range and small town, where sex role models were not challenged by urban fragmentation, but protected by the limited number of print media roles provided before the rise of movies and television.

Miss America is perpetuated in part by television commitments, and the Marlboro Cowboy is a product of a $22 million advertising campaign which has become "the most virile image in the advertising world. They (Marlboro Cowboys) have had more impact than the auto racers, the pro football players or burly truck drivers" (Los Angeles *Times*, 1975:14).

The masculine, rugged, tatooed, independent, and hardy images are meant to match the cigarette and models are paid $300 a day and appear on billboards and in newspaper, magazine, and other ads throughout the country, while Marlboro Country remains Western ranchland far from the urban pressures which desex and psychologically castrate the male and drive him to lung cancer in the polluted cities.

For the cameraman, the Sunbelt is rich with sports imagery and the pseudo-action of a spectator society. Sun and leisure help, and the winning team symbolizes a "big league" status for emerging cities which use sports imagery as a vehicle for civic rivalry and in the competition for tourists, new industry, and the piracy of old industry (like the abduction of the Brooklyn Dodgers and New York Giants by California). Sports superlatives flow freely across "objective" airwaves and sports desks, and national TV captures the background with new stadia and skylines, as civic wars are fought by the Cowboys, Broncos, Bucaneers, Rangers, Rams, Oilers, Astros, Comets, Nuggets, Suns, and Sun Devils.

The Sunbelt is merchandised by the romantic and nostalgic country and Western music, whose roots go back to the South: (blues, bluegrass, hillbilly, jazz, and rock). ("The South is Doin' It Again"—"The South and Its Music," *Country Music* magazine, March 1977). The longing for the past, the frustration of urban life, and the desire to escape its hostility through musical anesthesia provides a strong antiurban, antibig city theme to much of the Nashville Countrypolitan Sound which appeals to the urban transplant who can identify with Elvis Presley of the 1950s or Willie Nelson of the 1970s singing in big cities (Cook, 1973, Jenkins, 1973).

The romantic escape to a Sunbelt utopia is cultivated by the songs of the Sunbelt, from "California, Here I Come," "San Fernando Valley," and "Don't Fence Me In" of earlier Westward migrations to the nomadic escapism of the urbanized 1970s—"Galveston," "The City of New Orleans," and "Let's Go (Back to the Basics) to Luckenback, Texas." Like shifts in magazine rankings of cities, the songs of older Northeast cities are often tunes of yesterday like ("Shuffle Off to Buffalo," "Autumn in New York," "Meet Me in St. Louis") while the newer vagabonds sing of "Going Back to Houston," "By the Time I Get To Phoenix," and lyrically plead, "Come on L.A. and Let Me In," or at least "Do You Know the Way to San Jose" (Young, 1969).

Singers and songs and Sunbelt towns are often inseparable. Nashville ("Music City, U.S.A.") and Austin ("Nashville West") both use country-western music as both industry and image, with the newer "outlaw," progressive, and "redneck" rock emerging at Austin (Reid, 1974), where the Deep South and newer Southwest historically meet to blend rural and urban, past and present into a unified media image package (Young, 1976). The convergence of country singers and cities is evidenced when Harold Lloyd Jenkins, borrowing from Conway, Arkansas, and Twitty, Texas, changed his name to Conway Twitty (Oppenheimer, 1976); and Henry John Deutschendorf, Jr., who grew up in New Mexico, Texas, Oklahoma, and Arizona changed his name to John Denver (Curtis, 1974).

Denver not only celebrates Denver, the Rockies, and his Aspen retreat, but "The Sunshine Boy" (Orth, 1976) gushes freedom and fresh air across the Sunbelt to West Virginia ("almost heaven") pleading "Take Me Home, Country Roads" for the homesick exurbanite who imagines "Thank God, I'm A Country Boy!" Country music appeals because of displeasure with modern urban

America, and the lyrics often suggest that "the South has none of the problems of the North, and the country has none of the problems of the city," as "Goodness is concentrated in the South and in the countryside, while badness is far more common in cities and in the North" (Marsh, 1977:80-81).

The bad press of the South can be compensated for by country music in which attributes become virtues, as "backward easily becomes rustic, ignorant becomes simple and uncomplicated, and reactionary becomes oldfashioned." This same analysis postulates that country music has some relationship to the recent positive images of the South and that "neither the election of Jimmy Carter to the Presidency, nor those southward migrations of population, industry, and political power to the so-called Sunbelt, would have been possible in the face of strong anti-Southern sentiment" (Marsh, 1977:82).

But shadows and sunset may force the Sunbelt to face new music in Paradise Lost, where shortages of water and fuel threaten drought, fires, and cars with no air-conditioning, as over-sold "consumers" and civic rivalry from the Snowbelt force the media to face the snakes in Eden as well as the worms in the Big Apple. Since political conventions have moved to the Sunbelt–Los Angeles, Miami, San Diego (almost)–and four of the last six presidents were Sunbelters or consumers of sun pleasure (fishing, golf, sailing, swimming, skiing), perhaps future presidential press conferences, libraries, and birthplaces will decentralize Washington and the White House even closer to the national urban crisis at Warm Springs, Key West, Key Biscayne, Denison, Abilene, Johnson City, Austin, Yorba Linda, San Clemente, Plains, and St. Simons–where the newsmakers and urban problems have already arrived.

NOTE

1. The authors, titles, and dates of the New York *Times* series are as follows: R. Reinhold and J. Nordheimer, "Sunbelt Region Leads Nation in Growth of Population," February 8, 1976; J.P. Sterba, "Houston, as Energy Capital, Sets Pace in Sunbelt Boom"; W. King, "Federal Funds Pour Into Sunbelt States," p. 24, February, 9, 1976; R. Reed, "Sunbelt Still Stronghold of Conservatism in U.S.," February 10, 1976; R. Reed, "Migration Mixes a New Southern Blend," February 11, 1976; B. Drummond Ayres, Jr., "Developing Sunbelt Hopes to Avoid North's Mistakes," February 12, 1976; J.T. Wooten, "Aging Process Catches Up With Cities of the North," and W. King, "Fort Myers Pacing Nation in Growth," p. 14, February 13, 1976.

REFERENCES

ALLEN, S. (1974). "Regional lifestyles: Better homes and gardens. Southern living. Sunset." In Magazine profiles. Evanston, Ill.: Medill School of Journalism, Northwestern University.

AXTHELM, P. (1977). "A voice of the south." Newsweek, (January 31):25.

BROWN, R. (1972). "Beyond Arizona highways: A survey of state-aided travel/promotional magazines." Paper presented at annual meeting of Association for Education in Journalism, Southern Illinois University, August 22.

BURD, G. (1969a). "The mass media in urban society." Pp. 293-322 in H.J. Schmandt and W. Bloomberg, Jr. (eds.), The quality of urban life, Beverly Hills, Calif.: Sage.

--- (1969b). "Critic or civic booster role for press." Grassroots Editor, 10 (Sept.-Oct.):7-8+.

--- (1969c). "Urban press: Civic booster." New City, 8 ((Jan.-Feb.):13-18.

--- (1970). "Protecting the civic profile." Public Relations Journal, 26(January):6-10.

--- (1972). "The civic superlative: We're no. 1/The press as civic cheerleader." Twin Cities Journalism Review 1(April-May):3-5+.

--- (1974). "A new state 'city magazine': A report on Texas monthly." Paper presented at annual meeting of Association for Education in Journalism, San Diego State University, August 19.

--- (1977). "Urban symbolism and civic identity: Vision and values in architecture and communication." Paper presented at annual meeting of Western Social Science Assocation, Denver, April 23.

CLAY, G. (1973). Close-up: How to read the American city. New York: Praeger.

COOK, B. (1973). "A-pickin' 'n' a-changin': Country-bred bluegrass music conquers the city." National Observer, (October 6):24.

CRYER, B. (1976). "Texans go hunting new fuels." Austin American Statesman, (September 26):B1.

CURTIS, O. (1974). "John Denver: Colorado's own superstar." Denver Post's Empire Magazine, (March 10):10-13.

DAGENAIS, J. (1967). "Newspaper language as an active agent in the building of a frontier town." American Speech, 42(May):114-121.

DEVEREUX, S. (1976). "Boosters in the newsroom: The Jacksonville case." Columbia Journalism Review, (January-February):38-47.

EGERTON, J. (1974). The Americanization of dixie: The southernization of America. New York: Harper and Row.

ERNST, S. W. (1972). "Baseball or brickbats: A content analysis of community development." Journalism Quarterly, 49(spring):86-90.

FLINK, J.J. (1975). The car culture. Cambridge, Mass.: MIT Press.

FLIPPO, C. (1974). "Gushing over oil in Houston." MORE, (January):10-14.

FLOAN, H.R. (1958). The south in northern eyes, 1831-1861. Austin: University of Texas Press.

FOSTER, M.S. (1975). "The Model-T, the hard sell and Los Angeles's urban growth: The decentralization of Los Angeles during the 1920s." Pacific Historical Review, 44:459-484.

FREIDEL, F. (1963). "Boosters, intellectuals and the American city." Pps. 115-120 in O. Handlin and J. Buchard (eds.), The historian and the city. Cambridge, Mass.: MIT Press.

FRINK, D. (1977). "Sunbelt: Another name for money belt." Austin American-Statesman, (February 6):G1.

GIBSON, J., and GREENE, J. (1973). Power, petroleum, politics and the Houston dailies. Austin: University of Texas. (mimeo).

GOLAB, J. (1976). "Rocky mountain hype." Denver Magazine, (February 4):38-41.

GORDON, M. (1965). "The Chandlers of Los Angeles: The world of Otis, Norman and 'Buff'." Nieman Reports, (December):15-19.
GRADY, H.W. (1966). "The new south: 1886," (ed.) T.D. Clark. Pp. 463-476 in D.J. Boorstin (ed.), An American primer. Chicago: University of Chicago Press.
GREEN, F.M. (1972). The role of the yankee in the old south. Athens: University of Georgia Press.
HENRICKSON, P. (1975). "Films called southerns: This corn makes plenty of bread." National Observer, (October 25:1ff.).
JENKINS, P. (1973). "Music city U.S.A." Manchester Guardian. Reprinted in Houston Chronicle, (September 2):6-7.
KING, W. (1977). "Atlanta opening New York office in a move to attract new industry." New York Times, (February 11):A14.
KUHN, F. (1966). "Blighted areas of our press." Columbia Journalism Review, 5(summer): 5-10.
Los Angeles Times Service (1975). "A visit to Marlboro country." Milwaukee Journal, (December 15):14.
MARSH, B. (1977). "A rose-colored map." Harpers, 255(July):80-82.
MARTIN, B. (1977). "The selling of Plains, Ga." St. Petersburg, Fla., Times, (June 5):1E.
McGILL, R. (1963). The south and the southerner. Boston: Little, Brown.
MOORHEAD, J.D. (1976). "The south's growing economic clout." Christian Science Monitor, (September 1):14-15.
MORGAN, N. (1961). Westward tilt: The American west today. New York: Random House.
OPPENHEIMER, P.J. (1976). "Conway Twitty: 'You have to have lived to sing from the heart'." Family Weekly, (January 18):9-11.
ORTH, M. (1976). "The sunshine boy." Newsweek, (December 20): 60-68.
PHILLIPS, K.P. (1969). The emerging republican majority. New Rochelle: Arlington House.
POLLAK, R. (1967). "How to build a publishing empire." Newsweek, (January 2):41-45.
REID, J. (1974). The improbable rise of redneck rock. Austin: Heidelberg Press.
SALE, K. (1975). Power shift: The rise of the southern rim and its challenge to the eastern establishment. New York: Random House.
SESSER, S.N. (1970). "The fantasy world of travel sections." Columbia Journalism Review, 9(spring):44-47.
TEMPEST, R. (1976). "The hard sell helps sunbelt keep booming." Dallas Times-Herald, (June 20):1ff.
VAN HEATHERLY, V. (1977). "Bye, plain Plains, hello, tacky town." Houston Chronicle Sunday Supplement, Special Issue, (June 12):1t.
VANDIVER, F.E. (1975). The southwest: South or west? College station: Texas A & M University Press.
WORKMAN, W.D., Jr. (1960). The case for the south. New York: Devin-Adair.
YOUNG, C.D. (1976). Redneck rock–Austin's newest city symbol: A survey of the music and the print media. University of Texas. Unpublished manuscript.
YOUNG, E. (1969). "Sing a song of cities." Nation's Cities, (October):29-30.

6

Sunbelt Boosterism: The Politics of Postwar Growth and Annexation in San Antonio

ARNOLD FLEISCHMANN

□ AS THE EXODUS FROM CENTRAL CITIES ACCELERATED after World War II, many older urban areas discovered that they were increasingly surrounded by suburbs whose growth and development was beyond their control. Despite the interdependent nature of their regional economies, coherent planning and control was precluded by the fragmentation of the metropolitan political structures. Such political fragmentation was often advanced as one more symptom in the long litany of ills which characterize the contemporary urban crisis:

> The most pressing problem of local government in metropolitan areas may be stated quite simply. The bewildering multiplicity of small, piecemeal, duplicative, overlapping jurisdictions cannot cope with the staggering difficulties encountered in managing modern urban affairs. The fiscal effects of duplicative suburban separatism create great difficulty in provision of costly central city services benefitting the whole urbanized areas. If local governments are to function effectively in metropolitan areas, they must have sufficient size and authority to plan, administer, and provide significant financial support for solutions to area-wide problems. [Committee for Economic Development, 1966:44]

The most frequently profferred remedies involved plans to integrate the local jurisdictions. However, despite numerous attempts to promote city-county consolidation, regional organizations, and metropolitan units of government, most northern metropolitan areas are still characterized by suburban autonomy and central city decline. In contrast to the lack of coordination characteristic of metropolitan governance in the Northeast, Sunbelt cities have confronted urban decentralization with a policy tool available to few of their northern counterparts—the annexation of unincorporated fringe areas (Dye, 1964).

This paper examines the annexation policies of one of the Sunbelt's most dynamic urban actors—San Antonio, Texas—in order to highlight some of the major political and economic variables which influence the use of this tool. The analysis will proceed in two stages. First, I will present a thumbnail sketch of San Antonio's growth, with special emphasis on the post-1940 period. Following that, I will discuss the political coalitions promoting San Antonio's growth, the means used by these coalitions to maintain or expand their political and economic power, and the relationship between these pro-growth coalitions and San Antonio's local governmental structure.

URBAN DEVELOPMENT IN SAN ANTONIO

Since its founding in the early 18th century, San Antonio has played a prominent role in the political and economic development of the Southwest. In 1837 it became an incorporated municipality and by 1890 it had become the most populous city in Texas. It held this ranking, due primarily to the economic stimulation of agriculture and the U.S. Army, until 1930. The depression days of the 1930s represented a period of stagnation which only started to reverse itself with the introduction of a new round of military spending in the prewar days of 1941. The combination of a reinvigorated military economy and the WPA helped launch San Antonio on a 30-year period of unprecedented growth. The political history of San Antonio during this recent period of growth represents a sterling case of the rise of a Sunbelt city. The city's business and civic leaders had never been idle during past periods of growth and they have been exceedingly active since 1940. The history of their actions is a chronicle of the promotion of unrestrained growth through the design of a highly supportive governmental structure. The key elements in this 20th century brand of boosterism proved to be council-manager government, annexation, and a well-developed service economy.

San Antonio's wartime boosters predicated their planning on the assumption that the "center of population of the country is gradually shifting toward this region" (Picnot, 1942:11). San Antonio's boosters also put all their economic eggs in the basket of the burgeoning service economy. The City Council and the Chamber of Commerce sought to make San Antonio a major link in transcontinental and Latin American air travel. The San Antonio *Express,* in its 12-point "Platform for Greater San Antonio," advocated among other things a "city plan" and a "great resort hotel." In addition to the Chamber of Commerce's successful efforts to relocate Trinity University in San Antonio, the city also began efforts in the early 1940s to obtain a medical school, and in 1944, the Commissioners passed a unanimous resolution asking the University of Texas Board of Regents to move its medical school from Galveston to a site furnished by the City of San Antonio.

In 1942 the city created its first permanent zoning commission, and in 1944 the San Antonio Planning Board was established to study and promote postwar development. The Board had a $150,000 budget administered by a four-man finance committee and was chaired by one of the city's largest builders and land developers. In addition to water, sewage, and street projects, the Board pushed for construction of a large coliseum, development of a downtown right-of-way for the projected international highway, and expansion of the Municipal Airport. In view of such efforts, a 1943 economic study classified San Antonio as among the six cities in the United States "which grew most rapidly during the war with the best prospects of retaining their wartime growth" (Egloff and Hauser, 1943:17).

However, war's end left San Antonio's boosters with a number of structural obstacles blocking the fulfillment of this glowing forecast. First, the city had not expanded its boundaries to incorporate the extensive and often poorly planned growth which occurred during World War II in fringe areas outside the city limits. Second, many unincorporated suburbs could not furnish a full range of municipal services, and San Antonio did not provide these services outside its corporate limits. Third, the residents of a number of suburban areas began moves to incorporate new municipalities in Bexar County.

And finally, the city's ability to finance postwar projects through both taxing and bonding was limited by law to a percentage of the assessed valuation of property in San Antonio. To surmount these problems, the boosters turned to a tool available to most Texas cities since 1912–annexation. As a home rule city, San Antonio was well-equipped for this task since the Texas Constitution and state law allowed the city to annex any adjoining unincorporated area by ordinance, with or without the consent of the affected residents.

As it evolved in the 1940s and early 1950s, San Antonio's annexation policy had three stages of development:

(1) to annex any adjoining developments which petitioned for inclusion in the city limits (1940-1944);

(2) to contain or stymie the territorial expansion of other Bexar County municipalities (1944-1951); and

(3) to prevent the incorporation of new cities in Bexar County (post-1951).

THE POLITICS OF ANNEXATION

San Antonio's earliest annexations resulted from suburban demands for new or improved services. Committees were formed in many fringe areas, often with the assistance of local realtors, to study whether a group of residents should incorporate or seek to be annexed by the city. The most frequently cited benefits of annexation were fire and police protection; the improved health conditions associated with city sewers, water, and garbage collection; lower utility rates; cheaper fire insurance; and inclusion in the San Antonio Independent School District.

In the mid-1940s the city increased both the number and size of annexed areas, and shifted to a policy of suburban containment. In abandoning its passive attitude toward annexation, San Antonio fell into step with Texas's other large cities, which increasingly felt compelled "to grab huge blocks of land to protect their industries, assure future tax revenues, regulate housing and health standards, [and] provide room for expansion" (MacCorkle, 1965:25).

In August of 1944, the Commissioners passed four ordinances annexing over 6,000 acres. Among the areas annexed were several miles of streets which had served as a boundary between three northside incorporated suburbs and the unincorporated portions of

Bexar County. San Antonio previously bordered these suburbs only on their southern and western limits. By annexing these thoroughfares, the city encircled the cities of Alamo Heights, Terrill Hills, and Olmos Park, and thereby prevented them from annexing any new territory. This left San Antonio with only unincorporated land on its northern border.

The major advocate of this containment policy was Gus B. Mauerman, who served as mayor from 1943 to 1947. In reviewing his policy and his term of office, Mauerman said, "I kept on my toes and never let any new suburbs grow. I took them in before they had a chance to grow. That is not popular" (Balmos, 1959). During Mauerman's tenure, the city annexed almost 18,000 acres, thus increasing its area by 68%. Some opposition to this policy developed, and 11 of the 25 annexations during the Mauerman years were approved by a three-to-two vote of the Commissioners. At one point, one of the mayor's adversaries on the Commission asked, "When are you going to stop annexing territory?" to which Fire and Police Commission P.L. Anderson replied, "When the city stops growing." By 1945 such opposition had subsided and in the 1950s approval of annexation ordinances had become a routine matter to the City Council.

In 1951, the voters of San Antonio were asked to express their preferences on two matters pertaining to revision of the City Charter. Voters were first asked to vote "yes" or "no" on a proposition to create a charter revision commission. Second, pending the approval of this proposition, the electorate was to vote for candidates to fill the seats on the charter commission. Candidates for the charter commission were elected at-large to places which were designated only by number. This electoral arrangement produced a campaign between two slates, one advocating the existing commission form of city government, and the other supporting adoption of council-manager government.

Mayor White, the charter revision proposition, and the council-manager slate all scored easy election victories in 1951. The San Antonio *Express* left no doubt that boosterism had also triumphed in this election when it editorialized that the next order of business for the city was adoption of a "new Charter which will be simply phrased, lucid and workable, and which thereby will serve the interests of all the people—and keep San Antonio growing" (*San Antonio Express,* May 10, 1951:8).

The charter commission wasted little time in beginning its assigned task, and selected as chairman W.W. McAllister, a former Chamber of Commerce president, chairman of the city's largest savings and loan, and a future mayor. After two months of work, the charter commission submitted its report to the City Council and recommended that voters be asked to approve an entirely new document rather than a series of amendments to the existing charter. The Council-Manager Association, which McAllister had organized, poured $40,000 into the campaign to secure approval of the new charter. Like the earlier charter revision campaigns, this proposal was advocated in terms of its ability to improve efficiency, lower taxes, and remove politics from city government. In October, 1951, 30,000 San Antonio voters approved the new charter by a two-to-one margin, and the following month a slate of candidates selected by Mayor White was handily elected to the nine at-large seats which comprised the newly created City Council.

The new charter allowed the city to annex any unincorporated territory by a majority vote of the City Council. The charter commission could not have been unaware of the effects of annexation on San Antonio's growth when they included this procedure in the new charter. From 1940 to 1950, for example, San Antonio had increased its population by 29,500 through annexation. Migration added another 58,500 and natural increases an additional 66,400 residents to the city. This was in sharp contrast to the 1930s, when the city annexed no territory and net births comprised 77% of the total population growth.

In March, 1952, an ordinance was introduced which proposed the annexation of 120 square miles of territory, an area that was three times as large as all previous annexations, and which would have increased the city's population by more than 40,000. This proposal transformed San Antonio's previous policy of containing existing suburbs into one of preventing the incorporation of new municipalities in Bexar County. Following a March 10 meeting of the Planning and Zoning Commission, City Manager C.A. Harrell forwarded the commission's annexation recommendations to the council along with a memorandum in which he stated:

> From a standpoint of long-range planning and from the standpoint of the central city controlling the areas developed on its periphery with the resultant social problems and their impact upon the central city, it is the opinion of the City Manager that it is better for the City to take the

necessary steps at the present time to control this growth rather than to let it develop in a haphazard manner, hoping at some future date to remedy the situation.

Though not on its agenda, the proposed annexation was brought before the City Council on March 12. The ordinance was introduced at 5:00 p.m. after the council had completed action on its published agenda and after the final edition of the afternoon newspapers had reached San Antonio's streets.

A public furor followed the city's action, bolstered mainly by the opposition of Al Jergins and Strauder Nelson, two oil millionaires whose estates were included in the area proposed for annexation. A recall move against the council was begun, and in the face of such heated opposition, the city manager resigned and the council amended the original plan by reducing the area to be annexed by one-third. The council completed action on the amended ordinance in September, annexing 80 square miles and 32,000 citizens whose first tax bills would be payable April 1, 1954. Throughout the entire controversy, though, the City Council adhered to the guidelines which City Manger Harrell had offered as the basis for considering this new annexation policy:

(1) to protect the future of greater San Antonio;

(2) [to fight the incorporation of suburbs, which] destroy the growth and progress of any community, and are a bad thing for all concerned;

(3) San Antonio was already afflicted with five such parasite incorporated towns . . . [and it is] known that many more were on the verge of seeking incorporation, which would have completely encircled San Antonio; and

(4) [to catch certain areas off guard before the city found itself with] a full dozen or more parasite cities which would have lived off San Antonio for perpetuity without contributing a thing to its growth.

Boosterism suffered a setback in the 1953 elections, due in large part to the 1952 annexation. Three of the council incumbents declined to run for reelection, and Mayor White, who had abstained on the final annexation vote as part of his personal effort to eclipse the authority of the city manager, headed a ticket which was bankrolled by $100,000 from oilmen Jergins and Nelson. The Citizens Committee slate, which included the five remaining incumbents on the council, was no match for the White faction, which captured all nine seats on the council, two of them without a runoff.

In the wake of this defeat and Mayor White's efforts to establish himself as a strong mayor, Chamber of Commerce President Tom Powell formed a group in December, 1954 to plan a campaign to strengthen the council-manager system and to recapture City Hall in the 1955 elections. Drawing its membership from such earlier groups as the Council-Manager Association and the Citizen Committee, the 60 citizens at that meeting formed an organization known as the Good Government League (GGL). The GGL raised $40,000 for the 1955 election and fielded eight candidates, all of whom easily won council seats. Although San Antonio's elections are technically nonpartisan, the GGL quickly became the dominant force in San Antonio's politics, controlling every council seat it sought from 1955 to 1967, and maintaining majority control of the council until 1973.

The council, mindful of the fate of its 1952 predecessors, proceeded slowly with annexations in the early years of GGL control and concentrated its efforts instead on fiscal matters and public works improvements. Several new suburbs were created during this period, and by the end of the decade, the manager and council saw a need to once again prevent further incorporations in Bexar County. Previous annexations had illustrated the problems associated with taking large tracts of land into the city, the most serious of these being the need for San Antonio to provide municipal services to areas which added no immediate revenue to city coffers. This dilemma existed because property is required to be on the city's tax rolls (which are compiled annually from June 1 to August 31) for one year before taxes can be levied.

In order to overcome these difficulties, the city needed to develop a strategy which would allow it to avoid annexing large tracts of land while still preventing their incorporation or annexation by other Bexar County municipalities. The policy developed by the city manager and council relied on several provisions of the City Charter and Texas law. These required an annexation ordinance to submit to two readings before it could be passed. After the first reading, the affected area could not be annexed by another city nor could it incorporate as an independent municipality. With this in mind, a series of ordinances was introduced in September, 1959, calling for the annexation of 330 square miles. The council had no intention of annexing this territory but planned to prevent suburban incorporations by having the first reading and then indefinitely delaying final passage of the proposal. If this annexation ordinance had been enacted, San Antonio's population would have increased by 76,000 and the city's area would have grown to nearly 500 square miles, making it the largest city in terms of area in the United States.

Texas's large cities, especially Houston, resorted to this practice so extensively during the late 1950s that pressure was placed on the Texas Legislature to curb the annexation power of home rule cities. Most of the early opposition to this annexation strategy came from the Texas Farm Bureau and other agricultural groups. They were later joined by the Texas Homebuilders Association, which wanted builders to receive utility service while developing homesites outside city limits, and the Texas Manufacturers Association, which wanted to allow the creation of annexation-free industrial districts within a city's corporate limits. To resolve this dispute, the legislature finally passed a compromise measure, the Municipal Annexation Law of 1963, which had the following major provisions:

(1) Cities were granted the power to regulate subdivisions in an extraterritorial jurisdiction (ETJ) which extended to unincorporated land continguous to the city limits. The size of an ETJ depends upon a city's population. In the case of San Antonio, the ETJ extends five miles from the city limits.

(2) A city's ETJ is not subject to municipal taxation, nor may new municipalities be incorporated within the ETJ.

(3) A city can only annex land in its ETJ which is contiguous to its existing limits, and its total allowable annexation in any calendar year is not to exceed 10 percent of its area as of January 1 of that year.

(4) Any unused portion of a city's 10 percent allocation may be carried over to subsequent years so long as the total in any year does not exceed 30 percent.

(5) The 10 percent figure includes only land forcibly annexed, i.e., without the consent of a majority of the area's residents. The law further excluded from the allocation land used for a public purpose by any level of government.

(6) As a city extends its boundaries through annexation, its ETJ also expands.

(7) As the Texas Manufacturers Association had sought, a tax-exempt industrial district could be created in a city's ETJ for a period of up to seven years. This district is created by contract between the industry and the city, and is renewable for periods of up to seven years. The industry is thus exempt from city taxes but can be subject to zoning and other regulatory ordinances.

(8) Cities are required to hold public hearings on annexation ordinances, and must complete action on a proposed ordinance within ninety days after it is introduced. [MacCorkle, 1965:28-32]

San Antonio officials obviously found the act a mixed blessing. As one might expect, the major task confronting the city was the development of methods to preclude the incorporation of new suburbs. The solution developed by City Manager Jack Shelley was a policy that has come to be known as "spoke" or "finger" annexation. Under this program, the city would annex a major roadway within its ETJ while taking little, if any, of the land adjoining the right-of-way. This policy allowed the city to measure its ETJ from these spokes rather than from the old city limits, and thus, the city could substantially enlarge its ETJ without having to provide municipal services to large tracts of land.

In August, 1964, Shelley placed ordinances before the council which led to the annexation of nine spokes. With the exception of an abstention on one vote, each of the ordinances was unanimously approved. The total area taken in by these measures was only 1446.4 acres, far less than the 11,000-plus acres San Antonio could have annexed under the 10% allocation allowed by state law. Six of the spokes extended five miles from the city limits to the boundary of the 1963 ETJ. This action expanded the ETJ by an arc with a radius of five miles from the furthest point on the spoke. Hence the council effectively bypassed the state law which had sought to limit the quantity of land which a city could indefinitely reserve for future annexation.

Because of this expanded ETJ, the City Council no longer felt impelled to annex in large blocks and maintained a hands-off stance on annexation during the late-1960s. From 1967 through 1970, the city completed action on 56 annexations totaling about 1,300 acres, 55.5% of which (35 of the parcels) was annexed at the request of five of the city's major land developers and builders. Had the city chosen to, it could have annexed over 54,000 acres during that four-year period. Instead, development was allowed to proceed within the ETJ without the threat of annexation by the city. In fact, city policy during this period was to annex an area only when requested.

From this period through the early 1970s, the area's builders and land developers found it advantageous to develop tracts within the city's ETJ. Normally, a builder would acquire land with the ETJ and begin planning a subdivision. When his plans were complete he would submit them to the San Antonio Planning and Zoning Commission since the Municipal Annexation Law of 1963 gave cities the power to approve new developments within their ETJ's. When the builder received the Commission's preliminary approval, he would begin selling purchase options on his subdivision. Shortly thereafter, usually about one to four weeks before the Commission's final approval, the builder filed a petition with the City Clerk requesting that the subdivision be annexed. Since annexing an area can take two to three months, the builder would receive final approval from the Planning and Zoning Commission while his annexation petition was pending before the council, and would have his land annexed about two months after the commission's final action. Once this annexation had been approved by the council, the builder closed the sale of the lots and homes in his subdivision. This meant that at no time did the developer pay city property taxes on his land. He paid only the lower property tax levied by Bexar County, while the new homeowner received the first tax bill from the City of San Antonio 12 to 18 months after the land had been annexed. As might have been expected, this policy encouraged leapfrog growth since the tax break for the builder kept the selling price of a home in the ETJ lower than that of an identical home in San Antonio, and this lower price was translated into better mortgage terms for the home buyer in the ETJ.

In one respect, the city found this passive annexation policy beneficial. Once a developer received final approval for a subdivision, he posted a performance bond with the Planning and Zoning Commission guaranteeing that certain street and utility projects would be completed within a three-year period. Once these projects were completed to the satisfaction of the Public Works Department, the city refunded the bond to the developer and reimbursed him for the improvements. During the late 1960s, these bonds were usually posted before the public hearing on the builder's annexation petition. Since the Municipal Annexation Law allowed the residents of an annexed area to file a disannexation petition if they had not received services comparable to the rest of the city within three years after having been annexed, the city's use of these performance bonds precluded the possibility of an area being deannexed.

Strong opposition to the use of spoke annexation developed in the Texas Legislature during the late 1960s and most large cities in the state retreated from this policy. While the city maintained its passive attitude toward annexation in the late sixties, substantial development had occurred in the ETJ, and prospects seemed to indicate that rapid growth in the ETJ would continue. Therefore, in late 1970, the City Manager presented a proposal to the City Council which was, in effect, San Antonio's first effort to develop an annexation *plan.* Under the proposal, 155.8 square miles would have been annexed over an eight-year period. In accordance with the first phase of the plan, the council annexed over 4,000 acres in February, 1971. This included the site selected by the University of Texas Board of Regents for its San Antonio campus as well as extensive amounts of surrounding territory, much of it still rural. This annexation was located at the terminus of one of the spokes created in 1964, and, were it not for the spoke connecting it to the city, the campus would have been approximately two and one-half miles from the city limits of San Antonio. In fact, 18 months later it still took a council annexation of 14,000 acres to connect the campus site with the previously existing city limits.

City Manager Henckel was forced by the council to modify his plan following the 1971 annexations. The revisions called for a five-year timetable and included some areas which the manager had not proposed in his original plan. Full-scale political war over annexation erupted in March, 1972, when Henckel placed before the council the first stage of this revised plan. Henckel's proposal called for the forcible annexation of over 66 square miles, most of which was on the city's western and northern fringes. The area was to have been annexed on May 25 so that the property would be on the tax rolls before June 1, and therefore liable for city tax payments in 1973. Strong opposition to this proposal was led by residents of subdivisions near the Air Force bases on the city's southwest side. Resistance to annexation ranged from picketing city hall by the West Bexar County Citizens Committee to an extensive advertising campaign by the Committee for a Vote on Annexation, "which [was] openly underwritten by [builder] Ray Ellison" (Cook, 1972).

Ellison's action pointed to a cleavage that was developing among San Antonio's boosters. Some, like Ellison, favored a policy which would allow the area's land to be annexed only when a developer thought it desirable. Others, among whom were some builders, saw the regulation of San Antonio's growth as either necessary or inevitable.

On May 25, the council voted unanimously to annex over 63 square miles of unincorporated land. The following day, Ellison filed suit against the city and soon won his case because the area taken in by the city totaled 32% of San Antonio's January 1, 1972, territory. The city's action was therefore in violation of the 30% maximum imposed by the Municipal Annexation Law, and the council rescinded its action on June 15. Despite this setback, the council replaced the original annexation with a series of ordinances annexing 17 areas which had been included in the original plan. Although six citizens did appear before the council to oppose the overall plan, Ralph Langley appeared representing a proannexation group which he reminded the council was responsible for developing "more than 65 percent of the lots produced in San Antonio and Bexar County," and employing over 30,000 people. Langley argued for a "rational, logical, evenhanded annexation program that would deal with one and all in identical manner," and also proposed a revision in the city's utility policies, which he asserted increased the sale price of an average home by $400 (City Council minutes, October 25, 1972:18-19). Suggestions that the Council temper its proposal also came from the editorial pages of the local press:

> The *Light* has long been on record for a consolidated city-county government that would eliminate costly duplication.
>
> Until true consolidation comes, annexation fits into the pattern and certainly is a "must" for a growing San Antonio. The *Light* has consistently supported orderly annexation. We still do. Just how much and how soon are items for debate . . . the forward-thinking position would have to lean toward all areas in Bexar County becoming part of the city. [*San Antonio Light,* April 5, 1972:7D]

Thirteen of the ordinances were unopposed at the October 25 public hearing, and were approved unanimously by the council. Ray Ellison Industries did raise objection to one of the ordinances, and one of the handful of other opponents told the council:

> The real reason for the annexation is you are unable to manage your own City budget and want us to bail you out regardless of what happens to our financial situation. [City Council minutes, October 25, 1972:26]

With the exception of an abstention on one vote, the remaining four ordinances were also approved unanimously on December 14, 1972.

The massive 1972 annexations sounded the death knell for both the GGL and the pro-growth coalition which had dominanted the political scene since the early 1950s. The politics of unrestrained growth supported by a compliant city council were coming under attack from several directions and as a result, several new growth offensives were stymied by the newly emerging opposition. On the one hand, several business groups representing the new wealth in San Antonio objected to government regulation and interference with their land development plans. As such, they viewed annexation as an unwarranted involvement in their private affairs. On the other hand, several other groups felt that the city government had been too responsive to the growth advocates while ignoring the needs of the older sections of the city. They too wanted to slow the pace of annexation but for the purpose of concentrating city spending in the older areas of the city.

With the disintegration of the pro-growth coalition, the GGL was forced to relinquish its political domination of San Antonio. In the 1973 election, the GGL lost control of the City Council for the first time since 1955. They also failed to recoup their losses in the 1975 elections. In both elections the major opposition to GGL candidates was spearheaded by a group of influential businessmen who were angered by what they felt was the city's increasing concern with managing growth.

San Antonio's political quagmire was further complicated in 1976 when neighborhood and environmental groups demonstrated their muscle for the first time in a January referendum to control growth over the Edwards Underground Aquifer, the source of the city's water supply. The referendum was designed to rescind a decision made by the City Council in 1975 to grant a rezoning variance which would have allowed a shopping mall to be built over the Aquifer. The Aquifer Protection Association spent $3,000 in opposition to the rezoning while its opponents received substantial support from the local media and put up $24,000 in an effort to secure construction of the mall. The referendum reversed the council's action by a four-to-one margin, thereby weakening the pro-growth business coalition and some of the city's major builders while enhancing the political power of several environmental groups and West Side neighborhood organizations.

In the wake of this setback, San Antonio's boosters had to contend with a new participant in the local arena. In April, 1976, the

U.S. Department of Justice claimed that 13 of the 23 annexations completed during 1972-1974 violated the 1975 Voting Rights Act. Specifically, the Department of Justice charged that these annexations, coupled with an at-large method of electing the City Council, had effectively diluted the voting strength of the Mexican-American community. The Department of Justice offered San Antonio two options, either it could deannex these areas or switch to a district method of electing the City Council. After a good deal of wrangling, both in San Antonio and Washington, the Council agreed to hold an election in January, 1977, to amend the City Charter. The new plan proposed that each of the 10 council members be selected by district, and that the mayor, the council's 11th member and presiding officer, be chosen by all of the city's voters. In a bitterly contested election, the city's voters approved the new plan by a 31,530 to 29,857 margin, and ushered in what may be a new era in San Antonio politics.

CONCLUSIONS

Given these events and trends, how might we characterize postwar boosterism in San Antonio? The ideological foundations of San Antonio's politics can be characterized by a commitment to maximum growth, free of governmental interference. The city's boosters have viewed their relationship to local government as one in which the city is expected to be run like a private firm and provide those services which the business community considers necessary to preserve a healthy and growing economy. Consistent with this ideological base has been the commitment to "reform," which the boosters have characterized as a means of taking "politics" out of local government:

> The strength of council-manager government is the insulation it provides against wheeler-dealer council politics. And that strength depends upon the caliber of the manager's work. [*San Antonio Express,* April 11, 1976:44]

Unfortunately, such arguments pay little heed to the simple fact that no governmental structure is neutral. In their emphasis on governmental outputs rather than processes, San Antonio's boosters have denied that the reform institutions discourage or preclude political participation by certain classes, ethnic groups, or geographical areas. Perhaps Banfield and Wilson's description of another Sunbelt council-manager city, San Diego, is also applicable to San Antonio:

> Politics never intruded into the manager's office and seldom into the council chamber; to the casual eye, it might seem that the city had no politics. Actually, of course, important and controversial decisions were being made behind the scenes by leaders of the business community who exercised a controlling influence over both the council and the manager. [Banfield and Wilson, 1963:181]

By the mid-1970s, the boosterism of previous decades had begun to unravel, in part because of the birth of mass politics in San Antonio. Discord within the pro-growth coalition also developed as some boosters began to associate benefits *and* costs with the city's postwar growth patterns. During the previous 30 years, however, San Antonio boosterism had functioned within a unified framework of membership, organizations, and strategies.

In its early stages, the booster coalition drew its support from a coalition of business leaders, middle-class reformers, Republicans, and conservative Democrats who organized to oppose the Bexar County Democratic Party's New Deal orientation and ability to recruit and elect candidates in the city's nonpartisan elections (Randolph, 1973). During the postwar years, this coalition has promoted its objectives and mobilized its support by relying on trade associations, "public interest" organizations, and local political parties. Probably the major booster organizations in the local business community have been the Greater San Antonio Chamber of Commerce and the professional organizations representing the area's realtors and homebuilders, though occasionally ad hoc booster groups have been used for short-term goals such as promoting the international exposition held in San Antonio in 1968.

"Objective" examination of local public policy questions has been sponsored by the Governmental Research Bureau, which was active during the early 1950s when the city had a meager professional staff, and currently by the Research and Planning Council. Both of these organizations have obtained their financial backing and policy direction from local business leaders. Single-issue organizations, such as the Council Manager Association, have also used their "independent" status to promote booster goals.

The Citizens League and its successors, the Citizens Committee and the Good Government League, have been the formal political arms of this coalition and have drawn their financial support from the large contributions of local business leaders and the dues of members, most of whom have been residents of the affluent north side of the city. These local parties screened, endorsed, and financed candidates in the city's nominally nonpartisan elections. Before its dissolution in late 1976, for instance, the GGL was governed by a 36-member executive committee chosen by the organization's 2,000 members. This committee made major policy decisions and endorsed City Council candidates recommended by the party's candidate selection committee, which in 1975 screened over one hundred citizens interested in the GGL's nine endorsements and financial backing.

Annexation has proved to be an integral and consistent part of the booster platform. First, the fact that land was annexed almost exclusively in reaction to spatial growth, rather than to direct it, is a reflection of the traditional value placed on governmental nonintervention in the "free market." Second, just as the firm in classical economics seeks to maximize profits and control its marketplace, San Antonio attempted to eliminate competition arising from any other municipality in Bexar County. As Kenneth Jackson has observed regarding an earlier era:

> If counting the dead was frowned upon, annexing populous suburbs was a perfectly accepted method of fueling the municipal booster spirit . . . the thrust of municipal government was imperialistic, and the trend was clearly toward metropolitan government. [Jackson, 1972:448, 450]

Annexation has indeed made San Antonio markedly different from its Northern counterparts, both economically and politically. Between 1870 and 1970, for example, Boston's area expanded by only three square miles while San Francisco added only nine square miles between 1890 and 1970. San Antonio, on the other hand, added 110 square miles between 1950 and 1970. The decentralization which contributed in part to the breakdown of Northern pro-growth coalitions posed no threat to San Antonio, where annexation has allowed boosters to maintain would be suburbanites within their electoral coalition.

Without annexation, San Antonio would not be unlike San Francisco, Boston, or dozens of Northern central cities. For example, within its 1950 boundaries, the city lost 55,000 residents between 1960 and 1975. As a result of its ability to incorporate these outlying areas, the city's population, instead of declining, grew by 169,000 and the city was able to keep roughly 80% of Bexar County's population within its borders. The city's fiscal situation has also been enhanced by this territorial expansion. The 1952 annexation increased the city's assessed valuation by $141 million or 27.4%. This permitted a 12% reduction in the fiscal 1954 property tax rate, and by substantially reducing both the ratio of the city's net funded debt to assessed valuation as well as its net per capita debt, paved the way for a 1955 bond election. Thus, San Antonio's boosters have considered annexation a critical component of their strategy for growth:

> Annexation is an important growth policy for San Antonio. Because of that policy, followed for the past two decades, San Antonio has a relatively healthy city. Most of the other large cities have suffered economic-social strangulation by suburbs that drain away resources that starved the center city. [San Antonio *Express,* April 6, 1971:10A]

San Antonio's postwar growth has been rapid and sustained. Annexation and government restructuring have been key elements in the booster effort over the last three decades. That platform, however, has come under increasing fire recently as new groups have arisen to challenge the booster tradition and, in the process, make the city's political future uncertain. What types of growth will occur in this new political environment remains to be seen.

REFERENCES

BALMOS, D. (1959). "Trying times plagued 2 mayors." San Antonio Light, (September 8):13.

BANFIELD, E., and WILSON, J.Q. (1963). City politics. New York: Vintage Books.

Committee for Economic Development (1966). Modernizing local government. New York: Author.

COOK, J. (1972). "Big annexation set." San Antonio Express, (March 3):1.

DYE, T.R. (1964). "Urban political integration: Conditions associated with annexation in American cities." Midwest Journal of Political Science, 8:430-446.

EGLOFF, G., and HAUSER, P.M. (1943). Population shifts and postwar markets. New York: American Management Association.

JACKSON, K.T. (1972). "Metropolitan government versus suburban autonomy: Politics on the crabgrass frontier." Pp. 442-462 in K.T. Jackson and S.K. Schultz, Cities in American history. New York: Knopf.

MacCORKLE, S.A. (1965). Municipal annexation in Texas. Austin: University of Texas Press.

PICNOT, T.N. (1942). An economic and industrial survey of San Antonio, Texas. San Antonio: unpublished manuscript.

RANDOLPH, N. (1973). The citizens league will win again. San Antonio: Marjorie McGehee Randolph.

7

The Poverty of Public Services in the Land of Plenty: An Analysis and Interpretation

PETER A. LUPSHA
WILLIAM J. SIEMBIEDA

□ "WATER DEMAND PEAKS AFTER SUPPER in Shaw, Mississippi, when folks are washing up. In one part of town, the white part, water flows freely from the taps. But where the Black families live, it used to just trickle out of the spigots. There were bigger water mains and more water pressure in the white neighborhoods, and that is where one found most of the paved streets and the new street lights" (C.P.S.E.L., 1976).

One of the most widely held theories of urbanization is that given enough time, all urban areas (cities) will exhibit similar patterns of growth, stability, service provision, and decline. This theme, commonly called convergence theory, has been built up through many years of comparative analysis in the fields of sociology, economics, and geography.

Recently, however, this theory has been found wanting on at least two bases. First, the work of Brian Berry challenges the uniformity of urbanization thesis (Berry, 1973, 1975). Berry maintains that

AUTHORS' NOTE: *We wish to thank the following individuals for the thoughtful help they have provided: David Perry, Al Watkins, Charles Cottrell, Bob Anderson, Linda Bonnefoy, and Ruby Dannenberg.*

cultural differentiation and technological adaptation are the basic molding forces of urbanization. He contends that we should not expect all urban places to develop in a seminal manner, because of the varying means by which societies with general regional differences diffuse and adapt to mechanisms of developmental change. A second line of attack on convergence theory is provided by David Perry and Alfred Watkins (1977a). Their thesis states that there is no natural preordained process of urban development. What order occurs is directed by the combined investment efforts of the public and private sectors. As with any social planning policy that distributes scarce resources, shifts in spending and investment from North to South will alter not only the form of our national urban pattern, it will alter the impact of the urbanization process as well.

In this paper we throw our lot in the direction of Berry, Perry, and Watkins. We too challenge the conventional wisdom behind convergence theory and in this paper we will pay attention to the disparities in the provision and delivery of public services between the Southern and Northern regions of the United States. It is our thesis that the present gaps in the aggregate and particular levels of public services between these regions cannot be eliminated simply by bringing the Sunbelt up to economic and demographic par with the North. For, the differences between the regions are those based not so much on wealth, but on variances in cultural and political perspectives toward the public functions and responsibilities of their cities. In our view, the Sunbelt's lack of commitment to public services rests not only in the region's economic structure, but in its history and the evolution of its sociocultural and sociopolitical institutions and customs.

Differing developmental streams of political culture have made the political style and the public service performance of the Sunbelt region rather different from that of the urbanized Northeast. In his book, *The United States*, Ira Sharkansky recognized the Madisonian position regarding economic and developmental heterogeneity. He fails to recognize, however, that it is a heterogeneity of political cultures which has led to the differing levels and configurations of public service commitment and performance (Sharkansky, 1975).

THE DISPARITY QUESTION

A clear and accepted methodology for measuring public service disparities is still in the making. The present academic trend is to look at micro community or neighborhood level outcomes of distributional municipal resource policy (Levy et al., 1974; Kirlin and Erie, 1972; Lineberry, 1977; Cottrell, 1976). We agree with this approach but at present are forced to rely on secondary analysis and aggregate statistics. Of necessity, the output measures utilized in this paper are of the crudest form. For they are in reality input measures. Variables such as number of police per capita only give an approximate picture of the way in which the governmental decision-makers allocate their general resource base. They do not provide an accurate picture of what groups in society are receiving the services, at what level, and in which spatial areas (communities or neighborhoods). As Levy et al. (1974) have pointed out, a more objective output is one that represents the way the goods and services are supplied *and* received by the public. We believe that the only true output measures are those which show the amounts of services delivered to the polity by demographic subgroup, and the quality of what is delivered in terms of use. Such outputs are micro in nature, and represent real distributional benefits of public services in the equity sense. Such data are only available in the most fragmentary and subjective form. We therefore have, with these caveats, simply proceeded with what data we could gather. As our main agrument is theoretical and aimed at proving the sociocultural and sociopolitical bases of regional difference, we believe we can construct our framework with the paucity of data at hand. To the reader who proclaims we have left the icing off the cake, our response is that in this paper we are more concerned with the proper measurement of the basic ingredients. In future work we shall focus on the details. Here we only present the variables at the SMSA level grouped by size. As such, it establishes not only aggregate regional differentiation, but demonstrates that smaller cities in the Sunbelt exhibit even larger public service gaps than their larger brethren. After the macro comparison, we present a smaller, more humanly scaled picture of a particular Sunbelt city, San Antonio, which provides a finer grained picture from which to view the disparity question.

DO DISPARITIES EXIST?

The broad based public service indicators presented in Tables 1 and 2 demonstrate a few of the relative gaps that exist between the Southern and Northern rims. A more exhaustive empirical study disaggregating this data by subsector governmental unit responsibility would be useful in detailing the extent of service disparity. However, the absence of such data does not lessen the demonstrable nature of existing disparities.

In terms of per capita welfare expenditures, the Southern SMSAs in every size group do not expend even half the national average, and in some selected cases, the Northern cities are 3000% above their similarly sized Sunbelt counterparts (See Table 1). For municipal employment per one thousand population, the Southern rim cities in all but one case fall below the national average, and in no case exceed the Northern tier groups. Municipal employment is obviously a very important public service variable for it is a strong overall indicator of the polity's concern for public services. It is also a useful indirect measure of the pressures on the "demand" side for the improvement of public services.

Local public health expenditures are more encouraging for the Southern tier. Although falling just below the national average as a total group, they display marked improvement over the indicators discussed above (see Table 2). One possible explanation for this would take into account the different level of unionization between North and South. In almost every Northern SMSA, unions provide full-scale medical programs, thereby lessening the demand on the public sector for health services. The picture for per capita local revenue mirrors that of public health. The Sunbelt is still below the national average and the Northern group means.

For a picture of absolute service disparities, one must turn to case studies. The Hawkins versus the Town of Shaw, Mississippi, case is well known and well documented (US. 437F. 2nd 1286). The town was found to have systematically discriminated against black neighborhoods. The court ordered relief. In Texas, studies by Charles Cottrell and his students paint an equally depressing picture of San Antonio (Cottrell, 1976).

The following are excerpts from Cottrell (1976) describing San Antonio:

A study of three areas located in the Northwest, Southwest, and Southeast quadrants concluded that streets in the Southwest and Southeast quadrants (i.e. Chicano and Black neighborhoods) were more consistently dirty and were more consistently without curbs.

TABLE 1
LOCAL GOVERNMENT REVENUE PER CAPITA IN SELECTED SMSA'S IN THE SUNBELT AND IN THE NORTHERN TIER IN 1970

	Local Government Revenue Per Capita	Per Capita Local Expenditures Public Health	Per Capita Local Expenditures Public Welfare	Municipal Employment per 1,000 Population
U.S. Average	**$329.86**	**$2.96**	**$11.98**	**15.80**
SUNBELT CITIES				
Population 500,000 plus				
Houston, Tx.	$256.75	$2.70	$ 1.16	8.00
San Diego, Ca.	425.37	4.40	42.41	7.30
Phoenix, Ar.	394.02	3.53	.05	9.10
San Antonio, Tx.	262.64	2.30	.45	10.70
Jacksonville, Fla.	330.59	3.29	.30	11.60
Population 200,000-500,000				
El Paso,Tx.	223.05	2.62	.88	8.20
Tucson, Ar.	313.27	3.32	.16	8.70
West Palm Beach Fla.	402.19	1.49	6.85	14.60
Albuquerque, N.M.	287.17	1.70	.01	8.10
Population under 200,000				
Gainesville, Fla.	467.56	7.26	.32	14.90
Amarillo, Tx.	278.24	1.97	.63	10.20
Monroe, La.	250.31	1.74	.03	17.00
Tyler, Tx.	229.38	1.36	.89	9.00
NORTHERN TIER CITIES				
Population 500,000 plus				
Chicago, Ill.	340.32	2.51	9.09	12.40
Newark, N.J.	343.85	3.27	31.90	37.20
Columbus, Oh.	259.97	2.60	18.82	9.40
Cincinnati, Oh.	309.19	2.30	15.58	27.90
Rochester, N.Y.	426.09	8.23	32.50	33.40
Population 200,000-500,000				
Flint, Mich.	370.99	2.23	12.47	20.30
Bridgeport, Ct.	288.91	4.02	9.98	27.40
Worcester, Ma.	342.90	2.63	49.84	31.30
Trenton, N.J.	306.69	4.05	22.17	22.17
Population under 200,000				
Brockton, Ma.	325.87	2.10	50.25	23.90
Atlantic City, N.J.	337.60	3.49	46.15	46.30
Manchester, N.H.	194.62	1.64	2.96	21.40
Danbury, Conn.	506.90	5.60	37.29	21.60

SOURCE: Ben Chieh Liu, *Quality of Life Indicators in U.S. Metropolitan Areas* (1976) New York: Praeger & Co.

A public official's response to this condition was:

> The installation of street curbing is paid for by special assessments of the street residents. If they do not care, or cannot afford to pay, their area will have fewer curbs.

The Public Works Department has their bureaucratic answer:

> Our street sweeping equipment is designed to function effectively *only* where there are curbs for the brooms to "backboard" the debris into the machine.

In turn, such conditions can reinforce biases and old stereotypes as this typical letter to the editor suggests:

> The trash in the streets in the Southwestern part of our city is just a further example of the way our Mexican-American citizens do not care about their neighborhoods, much less our city.

Cottrell's (1976) detailed analysis concludes that, in large part, what lies behind the lack of services to minority poor neighborhoods is a distinct lack of political representation.

TABLE 2
GROUPED SMSA PUBLIC SERVICE INDICATORS-1970

	Comparative Public Welfare Expenditures	Comparative Public Health Expenditures	Total Municipal Employment Per 1,000 Population	Local Government Revenue Per Capita
U.S. Average	**$11.88**	**$2.96**	**15.80**	**$329.86**
SUNBELT CITIES				
Population 500,000+	$8.87	$3.24	9.34	$333.87
Population 200,000-500,000	1.98	2.28	9.90	306.42
Population under 200,000	.47	3.08	12.78	308.62
Average for Sunbelt N = 15	$3.77	$2.87	10.67	$316.30
NORTHERN TIER CITIES				
Population 500,000+	$21.58	$3.78	24.06	$335.88
Population 200,000-500,000	23.61	3.23	27.23	327.37
Population under 200,000	34.16	3.21	28.30	341.24
Average for Northern Tier Cities N = 15	$26.45	$3.41	26.53	$334.83

SOURCE: Table 1.

CONVERGENCE THEORY

Some have argued that the disparities between Northern and Southern public services will disappear over time. This position rests on the notion that a convergence of service levels will occur as wealth, productivity, technology, and urban population factors in the Southern Rim reach that of the Northern Rim. Further, the North, due to its earlier industrial and urban population, reached a high aggregate level of public services well before the South. The levels of services in the North flattened out due to a stabilization of the wealth, technology, and population index. The South, with a slower beginning, but a rapid rise in later years, is now catching up. It also levels out at some point as a function of the index. Or, to put it another way, the Southern Rim, as it modernized, will converge with the North with regard to service provision.

The problem with convergence theory rests in its absence of any direct recognition of behavioral, cultural, and political factors. An assumption of the theory is that rational economic choices and homogeneity of population function perfectly. The problem here, as with other simple models of the competitive marketplace, is that man is a social and political animal as well as an economic creature. The Southern rim does not possess the same cultural, behavioral, and political basis as the North. Just as tribal tendencies in various African countries prevent a functional integration of economic and political activities, so do the regional disparities between the North and South in the United States. To this we would add that the model overlooks the behavioral functions of experience, learning, and observation. Having seen the Northern tier, why should the Sunbelt migrants chose to emulate it?

One might expect convergence to occur if the structural conditions of the Sunbelt were like those of the Northern Rim. However, it is our view that they are not. We should not expect a set of public services similar to those of the Northern rim to occur in the Sunbelt because there are structural differentials not only in the nature of what produces wealth in the two regions, but also in the behavior of present Southerners and in the behavior and social norms of the new Southern migrant.

In some senses, the Sunbelt is the modern American frontier. It is the new area of settlement and, as such, possesses values differing from older settlement areas. The new migrants generally have higher levels of educational attainment than the national average. Many are older people who are leaving the North and taking with them their acquired wealth and their desires to maintain it. Their political outlook is conservative, viewing the experience of the Northern cities as resulting in poor living conditions, high taxes, and a deteriorating environment. As such, they will push for good quality basic services such as police, roads, and fire protection, but not heavy expenditures on social and human services. Lineberry (1977) notes that research on public service satisfaction indicates that middle-class citizens are happy with what they receive and do not desire increases in service levels. We believe that there will not be the political will of the populace to pursue high levels of public services.

One may question our assumptions on the nature of the political will of the new migrants. The argument would be one of the eastern migrant who is used to high service levels and sophisticated enough to place successful organized demands on local government. This argument will hold true to an extent. The eastern migrant will want at least something similar to what they left behind. But they will not want to recreate the experience of rapidly rising taxes, complicated bureaucracies, and paying for someone else's services. Their desire will be for simple adequate services. They will have adopted a conservative line toward the role of government. A pilot survey of urban migrants to one Sunbelt community finds that more than one-fourth of those coming from the Northern rim are lured by low taxes and economic advantages. While they might like services and shops as they had in the Northeast, they have no intention of taxing or creating governments to achieve them. Of those from the Northern rim, 55% do not desire to incorporate as a city; 74% do not want to be annexed to existing cities; 88% do not want any increase in taxes to improve services (Lupsha and Anderson, 1977).

The urban migrants are bringing a "privatism" to the Sunbelt—a privatism that should match rather comfortably the existing social norms. For example, as more of our senior citizens move to the Sunbelt, services particularly suited to their needs will increase—services such as emergency ambulances and paramedic care. We would expect, as with other advanced social services, that these will be provided first in the larger, wealthier Sunbelt cities and flow outwards to the smaller towns over an extended period of time.

Traditionalism is also strong in the Sunbelt. No matter how much wealth is accumulated in cities such as Houston and Atlanta, there is an implicit and instrumental traditionalism that is adhered to and, for the most part, desired. Call it heritage, call it parochial regionalism; it still exists. And it will continue to exist for some time. Southern folk have regained a pride that has been submerged for decades, and Northern innovation will be met with some skepticism.

The population and technology factors will also push the Sunbelt in the direction of service equality but probably more in the racial equity sense than in the market sense, as recent gains by Chicanos in San Antonio's City Council election suggests. The Sunbelt wealth is formed by the rise of corporate interests, however, in the mineral extraction, agribusiness, and technology areas. The benefits of this wealth have not been widely diffused throughout all segments of the population. The lack of unionism, right to work laws, a colonial "2-scale wage" system, and the disparities inherent in the slavery systems for blacks and the colonial system for Mexican Americans still have not been overcome. Thus we do not view wealth alone as the singular driving force behind Sunbelt convergence with the Northern region.

We must state again that the absence of a direct consideration of the differing evolution of political cultures places the convergence model in a weak position to explain public service gaps. It is our view that the differences in the evolution of political culture can be better explained by the following propositions:

(a) that the biological statement "ontogeny recapitualates phylogeny" has meaning for the study of political phenomenon in that all public policy is backward looking, rooted in the past. A glib way of saying this is "ideas have to be around for a while before they are accepted";

(b) that public policy reflects precedent behaviors that were efficient and meaningful at some time past;

(c) that public policy creates organizational structures which solidify that past into an ongoing,temporally moving dynamic reality;

(d) that need, primarily economic and physical survival need, is the mother of political and public policy inventions. But, that sociocultural and sociopolitical experiences shape the design.

(e) that age-development–life span analysis combined with an understanding of the above–provides an evolutionary perspective if used cross-generationally for understanding political events and phenomenon.

Using these guidelines we conclude that (1) our present is rooted in a past not of our making; (2) the political cultures do differ; (3) they rarely change without a diffusion and impaction of exogenous factors; and (4) our understanding of public policy can be enhanced through an examination of sociocultural and sociopolitical forces which underlie its creation.

LOCAL CONDITIONS FOR PUBLIC SERVICE PROVISION

Convergence theory tries to spell out a macro theory of service provision. On the micro, or city level, service provision is provided under a set of conditions which represent a "climate" for implementation. It is important to be fully aware of these climatic factors.

The first basic condition for the provision of public services rests in the values and attitudes of the dominant economic and sociopolitical elites. When urban elites desired rapid growth in their cities the public underwrote the cost of public services.

The second basic condition for service provision is the productivity or wealth of a given community. Wealth not only provides the tax base to meet demands, as well as respond to desires for amenities, it is also likely to increase citizen expectations. Here we make the assumption that as real per capita income rises, people demand (and expect) higher levels of public services, just as they make increased demands on the private sector for increased goods and commodities. We believe that the public service histories of cities like New York, Philadelphia, and Boston tend to bear this out. In most of the area now referred to as the Sunbelt, there has always been widespread poverty, underdevelopment, and, while there may have been need, few resources for an effective tax base. Recent growth in energy resource exploitation, agribusiness, and technology, as well as federal spending policies and inducements, have begun to alter this situation, but only with limited public service consequences.

The third condition for increased public service provision is simply increased urban population growth. The rapid expansion, in the late 19th century, of our Eastern cities brought about technological and political demands for increased public services. This connection between population growth and service demand gives rise, among convergence theorists, to the view that rising population in the Sunbelt will soon create service demands and social costing matrices quite similar to those of New York City, Newark, or Boston (Wade, 1976).

The fourth condition for the provision of public services is proximity. Our argument here rests with theories regarding the spread and diffusion of innovation. Close proximity to examples of public service provision allows for observation, adaptation, and diffusion of information and expectations. The dense and advanced patterns of urban settlement of the Northern megalopolises facilitated this process. And indeed, there are examples of local service provision that were quickly adapted at the statewide level throughout the Northeast. At the other extreme, the historical as well as geographic separation of urban centers in the Sunbelt did not provide a contagious environment for the rapid diffusion of urban service expectations.

The fifth condition for the provision of public services that is tied closely to any diffusion thesis is some relatively equal and relatively advanced level of politicization on the part of the citizenry. If the citizenry is not politicized for the purpose of making service demands, then obviously there will be little need for any elite response. It is our contention that this lack of politicization has been fairly commonplace in the Sunbelt, and we will develop this point in detail later on.

A sixth condition for the provision of public services is that the local political culture and community define such provision as a legitimate obligation and function of government. We argue that the fundamental difference in the disparate provision of public services in the Sunbelt, as compared to the Northern tier, is that elected political elites in the Sunbelt have traditionally believed that the provision of many common services—public transportation, emergency medical services, paved streets, curbing, etc.—went beyond the legitimate obligations and functions of the polity.

A seventh condition closely related to this is that the local public decision-making community possesses a shared view of the equity concept. As Levy, Meltsner, and Wildavsky (1974) have shown, a single community can possess very different views of equity, given different issue areas. In order for there to be the provision of public services, there must be some agreement on which equity model is to be expoused. It is our contention that the typical model for the political culture of the Sunbelt has been one of market equity. "You get what you pay for," for "Them that's got, get more." In San Antonio, Texas, for example, street curbing is paid for under a market equity model. So there are few curbs in poor—usually Chicano or black—neighborhoods.

The equity model has an analogue in the racial model. The racial model assumes that ethnics of color (blacks, Mexicans, Native Americans, etc.) receive access to public services only after whites receive them; and then only under the condition that all other demands on local government for use of public resources have been satisfied. Therefore, only after Anglo neighborhoods receive what they require, and local government has an implicit surplus, do allocations to ethnic neighborhoods occur.

Table 3 from Cottrell's study of San Antonio illustrates this last point, for greater local revenue sharing allocations were made to the richer Anglo neighborhoods (North and Eastside) than to poor Chicano and black (West and Southside) neighborhoods. Robert Lineberry's (1977) study of public service equity, which included San Antonio as a case, found little under class service disparity.

We, however, feel that Cottrell's data are persuasive and that in San Antonio the underclass service problem is present and real.

Unlike Sunbelt cities, organizational and institutional constraints tended to operate in the Northern megalopolises which gave reduced legitimacy to the market equity model and laid somewhat greater stress on models espousing equality of opportunity and, increasingly, equality of result.

An eighth, and our final, condition for the provision of public services is complexity. Namely, that when a certain level of

TABLE 3
LOCAL REVENUE SHARING ALLOCATIONS BY QUADRANT[a]

	Westside	Northside	Southside	Eastside
Number of projects	6	14	2	7
Expenditure level	2,416,745	3,761,794	650,000	2,825,000
Average median[b] family income	6,100	13,593	5,501	6,303
Average median home value	9,291	26,309	7,972	10,598

a. These differences exist in spite of the fact that barrio needs measured in terms of physical conditions would seem to require a different distributional pattern. In addition, when one examines project completion statuses by geographic quadrant (i.e., those projects funded through Local Revenue Sharing funds), there is a much higher completion rate among the Northside (Northwest and Northeast quadrants) projects. Hence, in spite of the apparent need for physical improvements and in spite of large wealth differentials, there is a significant inequality in the city council decisions to allocate Local Revenue Sharing funds. (Cottrell, 1976).

b. The data were drawn from only those census tracts in which projects were located.

SOURCE: Taken from the Ad Hoc Local Revenue Sharing Project, San Antonio, Texas, 1974.

complexity, density, cost and congestion is reached, labor intensive, individualistic solutions must be replaced with capital intensive, technological solutions. The latter can only be performed through the public decision-making of the polity. For many of the cities in the Sunbelt, this level of complexity has only recently been reached. For some of the smaller outlying areas of this region such levels are unlikely to be reached in this century. Some form of mass public transit was essential in pre-automobile cities like Boston, New York, Philadelphia, and Chicago and the public decision-makers in these cities have long recognized the need. Many Sunbelt cities are only beginning to consider the mass transit option. Indeed, one small Sunbelt city, Las Cruces, New Mexico, has even chosen to go the other way by abandoning its mass transit system because the city fathers have decided that transportation should be a private, not a public, responsibility.

THE HISTORICAL ROLE OF STRUCTURAL WEALTH DIFFERENTIALS

The disparity in public services between the Northern and Southern rims is based in its structural as well as its socicultural context. When we refer to the structural configuration, we are addressing factors relating to the historical production of capital and the overall economic configuration of the region. The Northern rim, historically, produced its wealth through industrial processes and industrial innovations.

The nature of industrial labor is one of a contractual basis for services. After the 1870s, fewer tool-owning craftsmen or artisans were found in the Northern rim and, instead, ever increasing numbers of factory and industrial workers dominated the labor force. The economies of scale essential for labor and industrial production demanded a density and a pooling of humanity in urbanized settings. Cities like Akron, Gary, Pittsburgh, Detroit, and Flint all grew in the industrial period providing habitat for pools of industrial labor (Perry and Watkins, 1977b). These industrial cities also provided an opportunity for politicization and organization both by the trade unions and by political machines. In short, the structure of the setting facilitated demand articulation and support mobilization.

The very nature of industrial work and urbanization means that any semblance of self-sufficiency must be achieved through cooperative activities, organizational activities, and, in combination with others, some active community of trade, bargaining, debate, and compromise. The worker relies on someone else to provide food, clothing, shelter, water, sanitation facilities, and security, in brief, all the essentials for an ordered existence of some duration. The essential needs of this type of community demand public action, public cooperation, and public services, because this is essential to the survival of the industrial polity and the generation of production and wealth. Thus, the very nature of the habitat in the industrial city requires an advanced level of public services.

In the Southern rim, wealth has traditionally been generated by agricultural and extractive pursuits. The regions did not develop in order to provide dwelling space for labor and the industrial machine. They exist to provide market centers, points of transshipment, and transportation nodes for the distribution of goods and services to and from the hinterlands (Nichols, 1964). The urban masses were not needed in the economic structuring of the Sunbelt cities in the 19th and early 20th centuries, for these cities served more as places of passage than as places of residence. Thus there was no need for a large or complexly organized urban population and, as a result, there was little need or structural demand for service provision. In the hinterlands, agriculturally based settlements had homogeneous and rather uniform needs. They tended to be individualistic, self-sufficient societies, in which the nature of the economy (agriculture, extraction) reinforced individual and small group activity. William H. Nichols in "Southern Traditions and Regional Economic Progress," documents the adverse effects of traditional agricultural thought on economic development in the South. He calls for a reevaluation of the objectives and goals of the Southern fathers for their home and its future. Historical anti-industrialization and mechanization modes of thought coupled with a desire to retain the traditional Southern way of life have impeded Southern progress and, he feels, will continue to do so (Nichols, 1964).

THE SOCIOCULTURAL DIMENSION

Because political ideas have to exist for some time before they are accepted in the majoritarian marketplace, the actions of today are often tied to the values of a previous era. Thus, the governmental outputs for the provision of public services in the Sunbelt today are embedded in the values of a political culture that may existentially no longer exist. In order to fully understand the disparity of public service provision that exists in the Sunbelt today, it is essential to understand its sociocultural fabric.

As Marx has so aptly pointed out, an area's social structure is shaped by its economy. The first sociocultural variable, then, is the Southern agrarian economy of the 19th century which required large, extended, and supportive family units. This was essential because labor was substituted for capital in the production process. Labor intensive agriculture required structures of communal cooperation and coercion. Cooperation on roads, fencings, wells, irrigation ditches, sowing, and harvesting was the rule—a rule that was maintained by private ties of kinship, affiliation, and affection. Similarly, merchants and bankers of the Sunbelt cities in the 19th and early 20th centuries had to cooperate with the regional economic structure and provide the capital and credit necessary to survive hard times. This cooperation promoted communalism but not socialism. For the glue of the affective bond was personal, a common need, and the structure it rested upon was land (Goodwyn, 1976).

Agrarian cooperation was reinforced by a second variable—fundamentalist religious conceptions and practice. It was a religion which preached individual effort, hard work, and self-sufficient responsibility. This fundamentalist perspective separated good from evil in crystal clear Calvinist dogma. There were the good and the bad; the unborn and the born again; those who labored in the vineyard of the Lord and those who were "no-accounts." These were crude distinctions, but vivid ones, and they took a physical form in the makeup of small town life in the Sunbelt. Being from the "wrong side of the tracks" or "Las Trackas," as the South Texas Chicanos refer to it, physically separates, in myth and fact, the God-fearing Christians from the no-goods. As we know historically, it also often demarked color, class, and ethnic separation and the line for the disparate provision of public services.

A third sociocultural factor in the low levels of Sunbelt services can be found in the personal backgrounds of those who settled the region. Most were second and third generation Americans. Most had never been exposed to foreign ideologies such as Catholicism, Judaism, Socialism, Syndicalism, or Communism. Few came to escape European despotism or persecution. Although many were Southerners uprooted and driven West by the Civil War, the collective experience of the Sunbelt's settlers was not one that looked toward government or organizations for the answers to problems.

More recently, several political scientists studying politics in Southern California have ascribed the rabid political conservatism of these Sunbelt residents to their previous history as bank- and government-shy Midwestern farmers and as ex-military officials (Wolfinger and Greenstein, 1969).

Thus, even today, we can point to agrarian roots, a strong sense of individualism, an overriding fundamentalist religious presence and a white in-migration of native born Anglo people as key ingredients in making up the sociocultural dimension of the Sunbelt—a dimension which, in combination with others, helps reinforce the pattern of low levels of public service.

In contrast, the sociocultural evolution of the cities in the Northern rim presents a very different picture. Northern cities evolved after the Civil War not only as industrial places, but as the domicile of foreigners. Jacob Riis, writing in 1900, noted:

> In New York . . . one may find for the asking an Italian, a German, a French, African, Spanish, Bohemian, Russina, Scandinavian, Jewish and Chinese Colony. Even the Arab, who peddles "holy earth from the Battery as a direct importation from Jersaleum." . . . The only thing you shall vainly ask for in the chief city of America is a distinctly American community. [Cordesco, 1968]

Common needs, common bonds of language, and ethnicity bound colonies of groups together. Unlike the "communalism" of the Sunbelt, on the Northern rim there evolved the "socialism" of the landless. The state, in the form of the city, had to provide the services that the individual self-sufficiency of the renter could not provide. The simple facts of urbanization, density, and the needs of industrial production demanded it, and these European and Slavic foreigners had the organizational skills and experiences to implement it. In short, it does not seem extreme to note that in terms of their sociocultural dimensions and evolution, the Southern rim and the Northern tier, though in one nation, were worlds apart.

POLITICAL DIMENSIONS

Sociocultural configurations must take on some organizational form if they are to affect public policy. A critical factor in the disparities in the provision of public services between the Northern rim and the Sunbelt rests on the different patterns of development and organizational forms taken by their local political systems. The critical elements shaping political forms in the Sunbelt were the values of privatism, individualism, religious fundamentalism, fiscal conservatism, laissez-faire, and a view of good government as good business. The "populism" essential to an agrarian economy was extended to the realm of politics. The structure and operation of local government was attuned to the needs of the land owners, not the residents of the cities (Goodwyn, 1976). In such a setting, local governmental functions were narrowly defined and the business community was actively recruited to govern the polity.

While Floyd Hunter's (1953) discovery of an economic power elite overseeing public decision-makers in Atlanta helped shock Northeastern academic pluralists into a decade long flurry of reaction, he was simply noting the political style of the Sunbelt. A traditional aspect of politics in the Sunbelt has always been the close relationship between the private economic community and the public decision-making community. In Dallas, the Citizen's Charter Association, a group of the city's top businessmen, have been instrumental in running the city since the 1930s. In Phoenix, a similar business group, the "Phoenix 40," have controlled local politics and questions of growth, land policy, and development. In San Antonio, 94% of all city councilmen elected since 1955 have been candidates of the Anglo business community's "Good Government League" (Gibson and Ashcroft, 1977). Similar situations exist in Alburquerque, Memphis, Miami, Tucson, and New Orleans where local business elites have had close working relationships and actively underwritten elected and appointed public decision-makers.

In addition to the close relationship between economic and political elites in the Sunbelt, one must keep in mind that nonpartisan politics is the rule. Partisan politics have always been taboo, but citizen organization politics–Good Government Leagues, Better Government Groups, Downtown Merchants Associations, Citizen Charter Associations, etc.–have long been a Sunbelt practice. It was, after all, in the Sunbelt that the ultimate form of government linking business to governmental administration–the commission system–was created. This was not simply a fluke of emergency, it was a political culture's natural reaction to crisis.

Nonpartisan government, whether the weak mayor-council form, the commission system, or city-manager government is usually somnambulistic government. For, it is a form of government that fosters a quiescent, acquiescent citizenry, where only the business community and the property-owning middle class need be politicized.

This lessened need for participation was reinforced by the election system common to the Sunbelt. Sunbelt cities are not only nonpartisan in form, they tend to be "at-large" in structure. This means that the potential organizational force of any ghettoized neighborhood ethnic or racial majority is dissipated and scattered across the entire community. As a result, electoral candidates and slates appeal to the broadest segment of the community, for those with the least sectional appeal have the highest probability of winning. In such a system, the weight of newspaper support and endorsement is greatly increased, for the local press is the critical medium of local political information for the limited attentive public. The at-large system of election contains an added element that makes it a perfect structure for system maintenance reflecting the pre-existing status quo—namely, under this system it is next to impossible to pinpoint specifically who is responsible for making decisions. Under this system, any request for policy must be tested against the will of the entire community, and any proposal that meets specific neighborhood needs no matter how beneficial can be rejected as lacking sufficient support. This becomes another factor in terms of public service equity. The which, where, and for whom of public service delivery in such a political culture is determined by that group which can mobilize effective campaigns to influence the local political will. In Sunbelt cities, that group has been the corporate business community.

Contrast this sociopolitical dimension of the Sunbelt cities with that of the Northeast and North Central regions and the effect of political culture on the provision of public services becomes even clearer. In the Northeast, political organization evolved out of the 19th century political machine. The political machine was the vehicle for control of city politics after the Civil War. The political machine has been analyzed in countless studies, so it suffices to say that the function of the machine was, in essence, the provision of rewards (jobs and services) in return for votes (Mandlebaum, 1964). What this meant, from our perspective, was that the political machine served a linkage function in Northern rim cities. It provided jobs, channeled dissent, organized ethnic groups, and provided public services for the masses, while working as a medium for capital accumulation by economic elites and maintaining the quiescence of potentially disruptive populations.

Through massive public works projects and service provision, the machine provided a public supported infrastructure for the enhancement of private investment and profit. The cost was patronage, graft and nepotism, resulting in massive public payrolls which are still evidenced in urban public employment patterns today. In addition, this form of organization meant increasing expectations for services and rewards in an ever spiralling manner. It worked smoothly as long as the propertied and investing classes or federal government largesse had a use and interest in the city. When they left, or their need for the city waned, the public sector was left with a pattern of reward, learning, expectation, and reinforcement that led to ever increasing demands for—even the notion of "entitlement to"—public services and few resources with which to fulfill them (Berry, 1975).

In contrast, in the Sunbelt cities, expectations were kept minimal. First, by the sociocultural evolution of the area and backgrounds of the population and, second, by a political organizational evolution that did not stress real or tangible rewards, but "community good"—that generalized notion of "best interest" that depoliticized those populations most susceptible to concrete, public service overtures.

To sum up our view of the political dimension, the disparity in public services between the Sunbelt cities and the Northern rim is, in part, a function of the differing evolution of their political structures and organizations. Political organization in the Sunbelt cities has never been designed to accommodate mass demands or create services. It was designed to function as an adjunct to the business and economic community providing a mechanism for accommodating growth and development. The differing nature of the organizational forms and the reward structure has led to a very different orientation toward participation and, in turn, the provision of public services. On the Northern rim, politics has been based on material rewards, while in the Sunbelt it tends to be rooted in symbolic appeals to generalized community values—like economic development.

A VIEW OF THE FUTURE

It is evident that the Sunbelt is changing. It is changing not only because of the massive influx of urban migrants, but because of sustained rises in wealth and the seemingly endless increases in its share of the federal budget. Service provision also is changing. We believe, however, that this change will take place within the ecology of the Sunbelt's traditional political environment. And, this environment is not likely to be amenable to a fundamental alteration in the present equity provision model. Lineberry (1977) aside, from all evidence it still appears that the underclass model has validity in the Sunbelt. If you are poor, in the future you may get something, but you will not get much. For the political culture and organization forms of the Sunbelt are not conducive to meeting material demands.

Large scale gains in initiating equitable service provision were begun during the great civil rights era of the 1960s. The positive enforcement of the constitution coupled with the acceptance of federal categorical grant programs that aided the poor and the ethnics, helped to remedy some of the abuses of the past, and bring others to light. The shift from categorical to block grant revenue sharing funding in our view reduces the likelihood of continued progress toward public service equity. For, this change in funding patterns fits perfectly into the historical evolution and political culture in the Sunbelt. It allows for more discretion, for more elite group influence, and for shifts away from the great society programming of the 1960s.

As we stated earlier in this paper, if you are to have a high level of public service delivery, certain conditions must be met. Historically, few of these conditions have been met in the Sunbelt. Today this is changing. Wealth, increased urban population, sense of proximity, density, and complexity are now everyday aspects of life in metropolitan centers like Dallas, Miami, and Houston. At the same time, other conditions—the values of organization and participation by underclass groups, a belief that a high level of public services is an obligation of government—or any widely agreed upon notion of equity have yet to gain a major foothold in the Sunbelt cities. Furthermore, the sociocultural and political organizational structure of the Sunbelt will not enhance the rapid implementation of these conditions. The Sunbelt cities are on the rise, but they will continue—as far as public service delivery is concerned—to pay far less attention to the needs of their poor and people of color than to the needs of monied and propertied elites.

It is time to recognize that the Sunbelt is different from the North. It has never been and, in our view, will never be a region where interclass coalitional politics and union influence will turn the political tide. The structural differences in economic base, culture, geography, and attitude mitigate against designing policy for the Sunbelt as one would for the North. The design of national policy and national resource programming must be done with a clear understanding of regional difference. One must be prepared to recognize the political culture for what it is, and what it desires to continue to be. Only in this way can those of us interested in equity and equality of public services hope to forge a strategy for improvement, if not outright change.

REFERENCES

BERRY, B. (1973). The human consequences of urbanization. New York: St. Martin's Press.

--- (1975). "The decline of the aging metropolis: Cultural bases and social process." Pp. 175-185 in G. Sternlieb and J. Hughes (eds.), Post industrial America: Metropolitan decline and interregional job shifts. Rutgers, New Brunswick, N.J.: The Center for Urban Policy Research.

BERRY, B., and COHEN, Y.S. (1973). "Decentralization of commerce and industry: The restructuring of metropolitan America." Pp. 431-455 in L.R. Massotti and J.R. Hadden (eds.), The urbanization of the suburbs. Beverly Hills, Calif: Sage.

CORDESCO, W. (1968). Jacob Riis revisited. New York: Anchor Books.

Conference on Public Service Equalization Litigation [CPSEL] (1976). "The next step: Toward equality of public service." New York: Trinity Grants Program.

COTTRELL, C.L. (1976). Municipal services equalization in San Antonio, Texas: Exploration in China Town. San Antonio: St. Mary's University Department of Urban Studies.

GIBSON, L.T., and ASHCROFT, R. (1977). "Political organizations in non-partisan election systems." Dallas: Southwest Political Science Association Meetings (paper).

GOODWYN, L. (1976). Democratic promise: The populist movement in America. New York: Oxford University Press.

HUNTER, F. (1953). Community power structures. Chapel Hill: University of North Carolina Press.

KIRLIN, J.J., and ERIE, S.P. (1972). "Development in research: The study of city governance and public policy making. Public Administration Review, 32:181-195.

LEVY, F.S., MELTSNER, A., and WILDAVSKY, A. (1974). Urban outcomes. Berkeley: University of California Press.

LINEBERRY, R.L. (1977). Equality and urban policy: The distribution of urban services. Beverly Hills, Calif.: Sage.

LUPSHA, P.A., and ANDERSON, R.U. (1977). "The lure of the sunbelt." Alburquerque: Working papers Division of Government Research.

MANDELBAUM, S. (1976). Boss Tweed's New York. New York: John Wiley.

NICHOLS, W.H. (1964). "Southern tradition and regional economic progress." Pp. 453-462 in J. Friedman and W. Alonso (eds.), Regional economics and planning. Cambridge: MIT Press.

PERRY, D.C., and WATKINS, A. (1977a). "To kill a city: A critical reevaluation of the status of yankee and cowboy cities." Studies in Politics: Series 1: Studies in Urban Political Economy, Paper no. 6. Austin: University of Texas.

--- (1977b). "Urban age and suburban development: A new method of dating cities." Paper no. 7. Austin: University of Texas.

SALE, K. (1975). Power shift: The rise of the southern rim and its challenge to the eastern establishment. New York: Random House.

SHARKANSKY, I. (1975). The United States: A study of a developing country. New York: David McKay.

WADE, R. (1976). "The end of the self sufficient city: New York's fiscal crisis in history." Pp. 1-4 in R. Wade (ed.), Urbanism past and present. Milwaukee: University of Wisconsin Press.

WOLFINGER, R., and GREENSTEIN, F. (1969). "Comparing political regions: The case of California." American Political Science Review, 63(1):74-85.

8

Economic Growth and Inequality, Demographic Change, and the Public Sector Response

ROBERT E. FIRESTINE

OVERVIEW

□ WHILE SOCIAL SCIENTISTS have had considerable time to study the general decline of the more "mature" cities of the industrial Northeast, they are just beginning the socioeconomic and fiscal evaluation of the "maturing" metropolises of the Sunbelt. As part of this effort, this paper seeks to examine some important aspects of economic growth and demographic change in the metropolitan areas of the region, with an eye toward the implications of these developments for the state and local provision of public services in the foreseeable future.

Scholars have generally confirmed the Kuznets hypothesis that the distribution of income becomes more equal as the economy "matures" (e.g., as income levels and other economic development proxies improve). This would auger well for anticipated future improvements in the distribution of income in the growing Sunbelt economy. However, diverse analyses of the relationship between socioeconomic characteristics and the degree of income inequality indicate that certain situations prevalent in the Sunbelt (of which race is the most crucial) may contribute to more–rather than less–income inequality as the metropolitan economy develops. To the extent that such factors remain potent in the future, concern for trends in income inequality as a policy problem may have to be placed much higher on the economic development agenda than has been the case heretofore.

Intimately linked to this concern is the issue of regional demographic change at the metropolitan and intrametropolitan level. We have only scant evidence of the changing character of the Sunbelt's population since 1970. However, from a wealth of earlier work in demography and migration, there is reason to believe that the composition of the continuing net in-migration to the region may have perverse effect on the intrametropolitan location of residential and nonresidential economic activity. This, in turn, could have serious consequences for the problem of inequality–both the economic inequality mentioned above and the related concern for fiscal inequality among local government jurisdications within individual Standard Metropolitan Statistical Areas (SMSAs). Briefly, it may be argued that much of the Sunbelt's new economic expansion relies on a relatively better-skilled, well-educated, higher-income labor force, the residential location preferences of which are generally in metropolitan suburbs rather than the central city. Intertwined with this is the continuing suburbanization of both production-related and consumption-related employment opportunities. Now, such migration-linked locational outcomes might well reinforce the region's seeming proclivities toward income inequality (in spite of the overall increases in income levels). Thus, the economic expansion of the Sunbelt could enhance the socioeconomic and fiscal standing of the metropolitan suburbs while leaving behind the less advantaged residents of the core cities (or, for that matter, the rural areas). Should this occur, the rise of the Sunbelt cities (read "metropolitan areas") could well produce the paradox of greater intrametropolitan socioeconomic differentiation between city and suburb in the face of enhanced levels of local and regional income. These possibilities further suggest a number of potential fiscal developments within Sunbelt local governments themselves.

ECONOMIC GROWTH

GROWTH AND THE CENTRAL CITIES OF THE SUNBELT

As perhaps the primary indicator of economic growth, recent increases in the level of income signify that the Sunbelt is rapidly taking its economic place with the rest of the nation. The region's high per capita income growth rate for the 1960s (7.5% annually, relative to 6.8% for the nation as a whole) continued into the present decade, though at a reduced differential (11.4% versus 11.1%, respectively, for 1970-1975, unadjusted for inflation).

With regard to socioeconomic change among and within Sunbelt metropolitan areas, it is useful to differentiate the South (arbitrarily, the states of the Old Confederacy lying east of the Mississippi, plus Kentucky, Arkansas, and Louisiana) from the Southwest (Oklahoma, Texas, New Mexico, and Arizona) and the West (Southern California and Southern Nevada). Other than the long-established identification of the South as a separate region, the two Western subregions are here distinguished from the South primarily because of their contrasting approaches to the problem of central city jurisdictional expansion. As the South has historically been more closely oriented toward the county as an important unit of local government, city-county consolidation has been the predominant vehicle for most redefinitions of the actual or functional scope of the metropolitan central city in that portion of the Sunbelt. The obvious examples of this are Nashville, Miami, Jacksonville, and Columbus. In the Southwest and West, on the other hand, central city annexation of surrounding territory has been the rule, as exemplified by many of the major central cities in Texas, Oklahoma, and Southern California.[1] Further, both geographic and historical circumstances serve to differentiate the more well established Western conurbations from their still maturing cousins in the Southwest.

This differential expansion of central city boundaries, relative to a much lower level of such activity elsewhere in the nation, is clearly crucial to any consideration of trends in intrametropolitan location in the Sunbelt. Unfortunately, there is little that one can really do to "adjust" for these changes; it is literally impossible to assess the changes which have occurred in a given central city by analyzing developments inside and outside its jurisdictional boundaries as they were defined in some previous base year. On the other hand, the evolutionary patterns in the subsequently annexed areas would surely have been different in the absence of central city geographic expansion. In any case, important socioeconomic and, particularly, fiscal data for the city proper exist only in aggregate form for the city as a whole. It is thus somewhat meaningless to ask how much the city spent during a given year within only a fraction of its current jurisdictional area (the "old" central city).[2] Therefore, in attempting to compare central city to suburb under such expansionary circumstances, about the best one can do in the aggregate sense is merely to be cautious in the interpretation of potentially misleading statistics. In the "overbounded" central cities (particularly in the Southwest), suburban communities have been subsumed by their aggrandized central cities. In the South, the generally more rigorous treatment of jurisdictional boundaries has led to such "underbounded" core cities as Atlanta or such county-delimited cities as New Orleans (Orleans Parish). In the West, contrarily, the metropolitan sprawl of Southern California has long made the construct of central city a statistical anachronism in the extended suburb which is the Los Angeles megalopolis. The caveat, then, is that the concepts of central city and suburb signify somewhat different entities within each of our Sunbelt subregions and in contrast to the remainder of the nation.

THE INTRAMETROPOLITAN LOCATIONAL EFFECTS OF GROWTH

From this, we would wish to examine the impact of regional economic growth on the intrametropolitan location of economic activity. Unfortunately, urban economics offers us few incontrovertible generalizations on the determinants of urban income (Conroy, 1975:39-63). While the greatest increases in income generally have occurred in those areas enjoying either predominant high wage industries or recent substantial growth of such industries, much of the Sunbelt remains dominated by low wage manufacturing (Conroy, 1975:57; Rees, forthcoming). Additionally, although relatively large proportions of employment in manufacturing are often positively associated with higher income levels (Mattila and Thompson, 1968; Perloff et al., 1960), Thompson (1975) found them to be negatively associated with recent population growth rates in metropolitan areas. Berry (1973:5), in fact, played down the role of the entire secondary sector (manufacturing and construction) in the growth of the metropolitan Sunbelt in favor of the tertiary (service) sector and the so-called quaternary (leisure-oriented) sector. At best, then, the relationships between economic growth in the region and the associated variations in employment composition and industry structure are not (and will not be) easy to sort out. Despite this, from our general knowledge of the process of urban economic decentralization, a few conclusions may be suggested about the impacts of economic growth upon intrametropolitan location decisions in the rising cities of the Sunbelt.

Manufacturing Activity. With respect to manufacturing, Rees's (forthcoming) analysis of the Dallas-Ft. Worth metroplex suggested that "suburban and exurban areas" recently experienced a greater influx of branch plants than did the central cities, though the latter enjoyed a relatively higher proportion of the establishment of new firms than did their surrounding areas. This appears to support the "incubator hypotheses" under which fledgling plants might first locate in obsolescent space within the central core until they are sufficiently strong to internalize some of their needed support services and thereupon move to more established facilities in outlying parts of the city or its suburbs. However, given the generally less intensive land use in many "maturing" Sunbelt central cities, the considerations of land cost and space utilization (which foster the decentralization of manufacturing) should remain relatively moderate in the Sunbelt for the intermediate future. Nonetheless, with reference to the rather low wage character of Sunbelt manufacturing, even spectacular success in attracting and retaining manufacturing activity to the central cities of the region would not be likely to produce significant advances in economic (income) growth per se. Nevertheless—at least in the intermediate term—it may well offer relative economic opportunity to those less skilled (and often racial/ethnic minority) residents of the core city who constitute the most endangered group with respect to the changing distribution of income in the metropolitan Sunbelt. Finally, regarding the appropriate strategies for sustaining a reasonable level of manufacturing activity within the central city, Struyk's advice is as relevant to expanding areas as to those on the decline:

> Upgrade public services in the industrial areas, aid with land acquisition for additions or parking lots, widen streets where possible to reduce congestion, and give the established firms tax incentives for expansion or modernization. If the central cities can substantially reduce the net out-migration of employment in this way, the pattern of new establishment locations is also likely to shift toward the more centralized locations. [Struyk, 1972:382]

Commercial Activity. As is quickly apparent from a glance at the downtown development plans of any Sunbelt metropolis, commercial space is of central importance in the expansion/rehabilitation effort. However, as Manners (1974) observed in his superb review, major portions of the office space which has recently been built in America's largest metropolitan areas—north and south—have been constructed in the suburbs. Significantly, he noted that the primary cost component in office operation is labor, not rent. Not surprisingly, Manners concluded that the often nonquantifiable and subjective judgments of senior management were "frequently accorded weighty consideration" in the decision to "go suburban." Thus, the largest and the smallest sized office firms—especially those relying on close physical proximity to clients—continued to prefer the central city. The flexibility of modern transportation and communications has enabled many formerly downtown-based operations to locate in the outlying areas, closer to both company production facilities and employee residences. While the evidence is unclear whether this sort of economic relocation has run its course in the nation as a whole, there is surely no danger of that in the Sunbelt. The relative importance of the costs of increasingly expensive labor are no less significant in the warmer climates, where the amenities of an abbreviated journey to work are not to be minimized. Without careful land use planning of the sometimes extensive open space that still exists in a number of the region's central cities, the corporate desire for new facilities in a growing commercial environment will certainly lead to the general enhancement of the suburbs with regard to this form of economic activity. Also, as will be argued below, the expected greater professionalization of the Sunbelt labor force could produce a substantial influx of such individuals into suburban residential (and hence employment) locations. In summary, as Manners implied, a form of accelerator mechanism may well be at work, where suburban office expansion itself becomes a function of continuing suburban growth rather than, as previously, simply a singular decentralization from the central city.

Retailing Activity. Briefly, much the same arguments apply to retailing, which remains more intimately linked to residential location patterns than is true for manufacturing or general commercial activities. On the one hand, to the presumed discomfort of central business district (CBD) retailers, the levels of urban retail activity have been positively associated with the growth rates of population and income and the relative prevalence of office activity, among other things (Berry, 1963; Friedman, 1973). On the other hand, decline in CBD business activity has also been ascribed to population decentralization outward to the urban fringe, evolution of the role of transportation in residential life-styles, and selective changes in age characteristics of local communities (Casparis, 1967; McDonald, 1975). Each of these items—obvious though they may be—reinforces the suggestion that the Sunbelt should prove no different than the rest of the nation in the inexorable suburbanization of retailing and other consumer-related economic activities. Indeed, as some of the growing core cities of the region lack the same tradition of "going downtown" to shop as may exist in more long established urban areas, retailing decentralization in the Sunbelt metropolis should ultimately surpass that of the nation as a whole.

Thus far, it would appear that there are indeed important concerns for the future location of economic (and residential) activity within the Sunbelt's growing SMSAs. With minor exceptions, many of the usual reasons for the suburban decentralization (and the "natural growth") of manufacturing, commercial, and retailing activity appear only to be intensified in the rush of new economic expansion throughout the region. Discomforting as these arguments may be for the future of interlocal economic and fiscal balance, there is yet a potentially more compelling concern—that of the evolving distribution of income within the metropolitan sphere.

INCOME DISTRIBUTION WITHIN THE SUNBELT METROPOLIS

In refutation of several of the more bleak suggestions presented above, the Sunbelt optimist would point with confidence to our earlier data on recent income expansion in the region, doubtless arguing that the continuation of such economic expansion would dwarf our fears for the future of the central city (even in its enlarged Southwestern form). Furthermore, he might argue, such projected growth would ensure to the lower status Sunbelt households a better opportunity for future income security than could be found elsewhere in the land: growth will alleviate the problem of poverty and with it, of course, the problem of discrimination.

Perhaps, and hopefully so. Nevertheless, there is troubling evidence from numerous studies of the changing patterns of income inequality that growth alone need not mitigate some otherwise regressive pecuniary effects arising from local institutional and socioeconomic circumstances. The most pertinent of these derive from problems of region-related racial composition of the population, the general trend toward urbanization, local industry mix, and the size and growth rates of metropolitan communities.

Directly, of course, many studies have demonstrated that income becomes more equally distributed as economic growth proceeds in relatively developed economies (Aigner and Heins, 1967; Al-Samarrie and Miller, 1967; Jonish and Kau, 1973; Ruthenberg and Stano, 1977; and Sale, 1974). This is in accordance with Kuznets's (1955) well-known hypothesis—which seems generally applicable to the modern Sunbelt—regarding the process of development in national (or regional) economies once a given level of systemic maturity has been attained. However, a more complex model of income inequality (as most often measured by the Lorenz Gini coefficient[3] arose from such additional considerations as the potential for racial discrimination (percent of population which is nonwhite), the relative level of population urbanization, region of the country, percentage of the labor force employed in manufacturing, and level of adult educational attainment (Soltow, 1960; and above). These state-level analyses clearly indicated that, even after income variations had been taken into account, greater inequality of income distribution was associated with higher proportions of nonwhite populations, more extensive urbanization, greater concentrations in manufacturing employment, and Southern regional location. (While *mean* educational attainment was, to be sure, linked with less income inequality, again the *distribution* of educational attainment would have been the more appropriate measure to examine[4]).

Such findings are reinforced by a variety of recent parallel studies of income distribution among U.S. SMSAs (Danziger, 1976; Farbman, 1975; Ford, 1977; Long et al., 1977). Again, these demonstrated the evident importance to relative inequality of both Southern regional location and (SMSA) population percent nonwhite, although Farbman (1975:235) did note some reduction in the explanatory power of the racial measure with the inclusion of the Southern regional dummy variable. Also, higher percentages in manufacturing appeared linked to greater inequality, as did SMSA population size and population growth rate.

Finally, noting the obvious implications of the above findings to the growth-related trends in the Sunbelt, several researchers have specifically turned their attention to parts of that region. Farbman

(1973) worked with 1960 county-level data for 12 individual states in or near the South (including Maryland and Texas). As confirmed by Rice and Sale (1975), he found race to be a generally more important explanatory measure of inequality than income on this interlocal level. Looking at four major national regions (South, Northeast, Midwest, and West) over time, Gunther and Leathers (1975) calculated that, while the decrease in regional income inequality during the 1950s was smallest in the South, that region showed the greatest relative decrease in inequality during the 1960s. Despite this significant change in trend, the South in 1970 still exhibited the highest level of inequality among the four specified regions. Also, when broken down by race, the results were not encouraging:

> The gains made in distributional equity during the 1960s were largely made by white Southerners. While the Gini [index of *in*equality] for the white families fell by over 12 percent, the nonwhite Gini declined by only 1.5 percent . . . [The parameters of these distributions suggest] that an upper class nonwhite group is emerging in the South, while the lower income classes remain relatively unaffected. This may reflect even greater poverty reductions among the white population than among the nonwhite. [Gunther and Leathers, 1975:21]

In summary, despite the dramatic income expansion in the Sunbelt in recent years, this economic growth has apparently done little to alleviate the relative extremes in income distribution which have marked the South, at least for decades. Moreover, many of the other institutional characteristics and socioeconomic tendencies accompanying the economic rise of the Sunbelt seem clearly linked to patterns of increasing—rather than declining—income inequality. High proportions of nonwhites (and most likely minorities in general, including Mexican Americans), expanding urbanization, greater population size, and higher growth rates in metropolitan areas, and even the relative importance of manufacturing (whose sustenance in Sunbelt central cities is advocated as a mechanism of employment for low-skilled worker/residents with few other job alternatives) —each of these has been shown to be associated with greater distributional inequality over and above the salutary effects of higher incomes. Assuredly, trends in a simple index of income inequality are inadequate indicators of an economically "better" or "worse" society (Taussig, 1977:70). Nonetheless, within the intrametropolitan context of this paper, these findings alert us to some possible undesirable economic consequences of the Sunbelt's urban heyday, particularly for many of its least advantaged, and often inner city, residents. To the extent that these fears are confirmed by the further research which is needed in this area, the outcome should be a more balanced approach to the comprehensive socioeconomic development of a maturing urban economy.

INTRAMETROPOLITAN RESIDENTIAL AND DEMOGRAPHIC CHANGE

RACE, RESIDENCE, AND INEQUALITY

To this point, we have argued that the accumulated knowledge of the past in certain areas provides a useful guide to the future: national experiences with economic growth, income distribution, intrametropolitan location of residential and economic activity, and the changing demographic milieu between city and suburb remain applicable even in the "new" Sunbelt. In moving from these issues to concerns for the interdependence among racial composition, income inequality, and residential segregation in Sunbelt SMSAs, however, our assessments must become more speculative in nature. This results from the somewhat fragmentary evidence on recent demographic change in the region and from the fact that such change itself appears to be without historical precedent.

Despite the very modest improvements in the *distribution* of income among Southern nonwhites during the 1960s (as noted above), the mean levels of Southern black income relative to those of whites improved significantly during that period (Shin, 1975). Farley (1977) reported that this was true for the nation as a whole during the 1960s. Moreover, he indicated that related black gains in educational attainment, occupational status, and employment opportunities also survived the economic recession of the early 1970s. While encouraging news at the macro level, these developments do not necessarily signify universal social improvements for urban blacks (or other minorities) in the Sunbelt. Indeed, from a survey of 32 Southern cities, Roof et al. (1976:59) concluded that "changes in socioeconomic status for blacks in the South have not been accompanied by concomitant changes in residential segregation." In comparing 1960 to 1970, the authors further noted that

> the overall evidence appears less persuasive now than a decade ago that socioeconomic factors play a large part in reducing segregation. This is not to suggest that economic factors affecting segregation are unimportant, or potentially insignificant. But it does suggest that economic and residential aspects of race relations are not very intimately linked. Black economic progress has occurred in the past decade, and continues to occur in the seventies, but without significant changes in residential patterns. [Roof et al., 1976:69]

This conclusion is supported by several recent findings pertinent to the Sunbelt itself. Working from census tract data for the 137 SMSAs which were fully tracted in both 1960 and 1970, Van Valey et al. (1977:839-840) concluded that their data provided "no support . . . for widespread downward trends in residential segregation" during the recent decade. Furthermore, in a study of both blacks *and* Mexican Americans in 40 of the 46 SMSAs in 1970, Frisbie and Niedert (1977:1007) reported minority income levels to be inversely related to the relative size of minority populations while "disparities between majority and minority income and occupation tend to grow as relative minority size increases." This reference to minority population size harkens back to our earlier observation of the link between population size and overall income inequality, and it accords also with the conclusion of Roof et al. that, within their 32 Southern cities,

> the overwhelming effect of city size is to increase residential segregation, not to reduce it by means of better jobs and salaries for blacks. This means that larger Southern cities are more segregated in 1970 despite improvements in the socioeconomic standing of blacks relative to whites.[5] [Roof et al., 1976:68]

Though admittedly fragmentary in nature, this evidence of a socioeconomically degenerative interaction among racial composition, income distribution, residential segregation, and urban population size—in the Sunbelt as well as the nation—is especially troubling in that the traditional palliative—economic growth—does not seem to relieve the problem. Indeed, given the particular racial/ethnic circumstances of the Sunbelt, the conditions of growth may well exacerbate these difficulties.

THE ROLE OF MIGRATION

Nor does the situation appear to be aided by the process of demographic change through population migration. It has long been recognized that interregional and intermetropolitan migration is selective of the younger, better-educated, and generally higher-income population (Weinstein and Firestine, forthcoming, Chapter 3), and there is increasing evidence (Adams and Nestel, 1976; Long, 1974; Long and Hansen, 1977; Weiss and Williamson, 1972) that this applies as well to black migrants to and between metropolitan areas. However, of course, intrametropolitan residential choice is severely constrained by household racial and income characteristics: implicit (though perhaps not demonstrably "illegal") discrimination confines those of ethnic minority (Hermalin and Farley, 1973) and low economic status (Harrison, 1974) to the metropolitan inner core. Additionally, the migration of upper- and middle-class whites to the suburbs—and their continued preferences for the suburban residential location—has been shown to be reinforced by the incidence of poverty in the central cities and by lower estimated net dollar values of central city budgetary expenditures to a middle-class family (Bradford and Kelejian, 1975). Long (1975) has identified such white out-migration *and* higher natural increase among central city blacks as contributing the most nationally to the increase in percent black in central cities. Black return migration to the South may well be of greater significance to that subregion's central cities, nonetheless, as that subregion attracts many of its expatriot blacks back to their Deep South areas of origin in the recent reversal of the traditional northward migratory pattern (Greenwood, 1976; Persky and Kain, 1970; Raymond, 1972).

A SUMMARY TO THIS POINT

The socioeconomic aspects of the argument are now essentially complete. It has been indicated that, both economically and residentially, a perhaps small though nevertheless significant proportion of the metropolitan population is largely denied access to the benefits which accumulate through the process of economic growth. For a variety of reasons, it seems quite possible that certain aspects of this situation may even be more troublesome in the Sunbelt than elsewhere (although these arguments are by no means definitive). Such a conclusion derives primarily from the particular confluence of racial composition, income distribution, residential segregation, population size and expansion, and demographic immigration which is peculiar to the metropolitan Sunbelt at this point in its history.

Now, certainly all the indicators of socioeconomic change in the region do not signal rough weather ahead. Nonetheless, the evidence is surely sufficient to suggest that intrametropolitan socioeconomic inequities in the Sunbelt may well increase, not decrease, with future economic expansion. In fact, it is difficult to imagine that evidence for a contrary thesis could be assembled in the amounts presented here. Accordingly, the rather pessimistic tone would not appear unjustified.

From this, then, what implications arise for the foreseeable response of the public sector to the changing character of the metropolitan Sunbelt: What are some of the fiscal pressures that might be expected to arise as a result of the new prosperity, and what problems might they pose for the societal issues identified here?

THE PUBLIC SECTOR RESPONSE

SOME TENTATIVE HYPOTHESES

With the continued economic expansion of the region, the positive income elasticity of demand for local public services should lead to expanded levels of local government taxing and spending. This may become even further enhanced as the growing proportions of metropolitan residents from higher spending regions begin to incorporate their tastes for (previously higher) levels of public service delivery into the local political spectrum.[6] Beyond this, Kelley's (1976) approach may properly be applied to the metropolitan Sunbelt, whereby increased government fiscal activity should derive from the growth-related characteristics of younger population age distributions, increased urbanization, expanded metropolitan populations, and higher population growth rates.

Of more particular interest, perhaps, are the expectations for local functional expenditures, especially as these may generally be differentiable between inner city and surrounding suburbs. Education, of course, is the most prominent local government function, with elementary and secondary schooling undergoing increased emphasis as the region seeks to divest itself of its historic image as a provider of inferior public education. In general, the trend of declining enrollments across the nation will be forestalled the longest in the burgeoning suburbs of the Sunbelt, where the continuing in-migration of children of the established middle (and, to an extent, upper) classes push higher both the academic and the tax/expenditure norms of earlier years. Coupled with (a) the explicit white flight to the suburbs (and its implicit version of immediate settlement in suburbia by in-migrants who otherwise would have located in the central city) *and* (b) the continued attraction of private schools for wealthy central city families, this growth will surely lead to decreased public school enrollments in the inner cities. Based on the evidence of intrametropolitan residential segregation and economic inequality, central city public schools may be expected to experience steadily rising proportions of pupils who are economically or educationally disadvantaged and/or who are members of national (though, perhaps, no longer local) racial/ethnic minorities (Goettel and Firestine, 1975).[7] Thus, expanding state government recognition of the increasing need for compensatory education efforts, plus continued federal funding of comparability-based ESEA Title I programs and related activities, should combine with declining enrollments to drive up per pupil expenditures substantially in central city school districts in the region. It is here, however, where the cost/quality expenditure argument (see note 6) becomes most sticky, for there results a numbing double paradox. First, the inevitable deterioration of the educational "milieu" as middle-class whites flee the central city public schools makes much more challenging the task of providing meaningful education to the increasingly disadvantaged and difficult student body which remains. However, as just pointed out, per pupil expenditures—even from nonfederal sources—will likely rise. Second, with this increased level of per pupil spending (often relative to many of the surrounding suburbs), the frequently wealthier central city districts (as measured by taxable property value per pupil) appear differentially advantaged. As a result, the natural suburban political strength is able to push even harder for revised state education aid formulas which provide increased support to the suburbs at the expense of the inner cities.

Though heavily financed by most states and accordingly linked to overall educational planning within states, public community colleges must nevertheless be considered here because of their intimate interaction with many of the issues addressed in this paper. Additionally, the public community colleges constitute that unique element of public higher education that is administered (and often financed in part) at the local level. With the growth of the region, demand for this flexible instrument of public postsecondary education may be expected to increase at a rather substantial rate, at least during the next few years. This is especially true in the metropolitan areas, where most new community college expansions will probably occur in the suburbs for the usual reasons of lower costs for large lot sites, greater planning flexibility in the design of physical facilities, improved private vehicle access (and parking) for predominantly commuter campuses, and a closer residential proximity for most of the potential student body—reasons which parallel those for the suburbanization of many types of private sector economic activity as well.

Unfortunately, such developments—while otherwise commendable—will likely work to the detriment of the less advantaged residents of the inner city whose need for the social mobility which education is commonly alleged to provide is surely no less than that of their more middle-class suburban counterparts. For these lower income and/or inner city citizens, however, their access to public community colleges (both physically and economically) as well as their career aspirations upon graduation are constrained by the circumstances described herein. As with elementary and secondary education, the less advantaged students in public community colleges are in need of instructional programs tailored to their particular situations, but this brings about a dilemma for the educational planners. On the one hand, they will be hard pressed to justify as rich a university "transfer" curriculum in an inner city community college as is often provided in suburban locations. On a per student basis, the generally larger enrollments in such programs at suburban campuses should prove more cost-effective as a result of economies of scale. On the other hand, decision makers may strongly restrict such "transfer" offerings in the inner city in favor of the more vocationally oriented "terminal" curricula, which themselves may be more costly as a result of greater requirements for facilities and equipment. If this is done, however, allegations of economic and racial bias should not be unexpected. The point is, of course, that both types of curricula are often essential at downtown campuses of public community colleges if all the potential students are properly to be served. If fully pursued, such a course must inevitably result in visibly higher per students outlays in such location. Naturally, such an alternative is economically and politically problematical in light of the increasingly middle-class nature of the Sunbelt metropolis.

Turning from education to the remaining major independent activities of local government,[8] the functions of public safety (police and fire protection) and of housing seem differentiable from the other "common functions" (Bahl, 1969) of sanitation and sewerage, libraries, parks and recreation, financial administration and general control, debt service, and miscellaneous. From Beaton (1974) and Chapman (1976), it is reasonable to expect increases in public safety spending throughout the Sunbelt metropolis, doubtless as a result of growing concern for the protection of increasingly valuable property and possessions. Indeed, this author found that, of all the major functional items of state-local spending examined in a recent study (Weinstein and Firestine, forthcoming, Chapter 4), outlays on police and fire protection were the most strongly affected by a variety of measures of migrant characteristics and population growth and change. Public housing, on the other hand, will doubtless remain strongly segregated—with its lower income residents—in the core cities, just as throughout the country. Finally, in response to local circumstances, tradeoffs may be expected to occur among the remaining functions, particularly across the very diverse set of central cities in the region. In the Sunbelt suburbs, on the other hand, nearly all the common functions should expand in response to income and taste related public service demands of the new residents. With this, of course, must come the inevitable increases in levels of local (and state) taxes to support public sector expansion.

PARTICULAR PROBLEMS

In closing, two aspects of the inequality dilemma seem of paramount concern. The first has to do with the implications of individual inequality as it may exist in the region at present, and the second relates to the potential for an increased level of intergovernmental inequality that may await parts of the metropolitan Sunbelt in the not-too-distant future.

The least advantaged in the metropolitan populations will likely remain in largely discouraging circumstances which may improve more by chance than by design. The cycle of income inequality, residential segregation, and decreased access to economic opportunity could become, for some, more vicious than in the past. Only a conscious concern for these issues on the part of development planners in both the private and the public sector can avert this unfortunate prospect. Moreover, it is not unlikely that—in all but the most extensive central city jurisdictions—the socioeconomic contrasts within the metropolitan population will produce increased fiscal (as well as economic) disparities between central city and suburb. Much has been done lately to assess this problem in the declining regions of the nation. However, too little effort has been expended to date in comparable inquiries for the still maturing metropolises of the Sunbelt. To the extent that Sunbelt central cities and, for that matter, some of their surrounding suburbs may reach a pause in the economic development process, these same sorts of fiscal ills which are becoming rather commonplace in the industrial Northeast may well make themselves more evident in the Sunbelt as well.

NOTES

1. Nonetheless, recent major annexations by Memphis and Knoxville suggest that our taxonomy is less than perfect.

2. Of course, in certain city-county consolidations—as opposed to annexations—the former central city may often retain some of its original functional responsibilities and thereby some operational identity.

3. The Lorenz Gini coefficient is a single statistic designed to express the degree of inequality of any given distribution across the specified (household) population within a defined analytical unit. Having ranked the households within a given jurisdiction (such as an SMSA) by annual income level (from lowest to highest), the cumulative percents of these households are then plotted on the X-axis against their respective cumulative percents of total SMSA income on the Y-axis. Complete distributional equality would be represented by a straight 45-degree line, wherein the lowest 20% of the households receive 20% of the income, etc. The Gini coefficient itself is a ratio of (a) the area beneath the 45-degree line but also above the plotted curve to (b) the entire area beneath the 45-degree line. Complete equality of distribution would yield a Gini of 0.0, and complete inequality a Gini of 1.0.

4. Educational attainment is not unrelated to the other determinants of inequality and thereby should perhaps be estimated in a separate structural equation. The results of Ruthenberg and Stano (1977) in just such an exercise, however, were not materially at variance with those of the more straightforward previous work.

Also, regarding the proper "cohorts" for an analysis of inequality, Paglin (1975) suggested an age-specific adjustment to the Lorenz Gini of income so as to avoid the alleged bias introduced by differing age distributions of households within different localities. This may do no more than beg the question of how properly to identify the numerous characteristics of family life cycle, as pointed out in Taussig's (1977:58-63) excellent summary of Paglin and his many critics.

5. Again, mean values mask distributional characteristics.

6. Even with this prediction of greater public sector activity, however, comes the thorny interrelation of costs (based on scale economies, local supply circumstances, etc.),

expenditures (derived in theory from both the supply and the demand for local public goods), and, for that matter, productivity (related to the efficient utilization of functional inputs to achieve the appropriate programmatic objectives). As generally suggested by Bahl (1968) and as demonstrated by Crowley (1970) and by Hamilton and Reid (1977) with regard to migration, the empirical differentiation between per capita expenditures on the one hand and unit costs on the other can confound the most careful and sophisticated researcher. Also, despite the recent emphasis on improved productivity in the public sector, we may know little more now than we did previously about the relationship between public expenditure (or public employment) levels and the real magnitudes of public sector outputs.

7. For a variety of supporting studies relating public education expenditures and enrollment levels, see Aaron (1969), Booms and Hu (1973), Greene (1977), and Gustman and Pidot (1973).

8. Nothing is said here about local spending for highways or for public welfare, as these activities are heavily dependent—financially and/or programmatically—upon higher levels of government. Health and hospital activities, also, are complicated first by their intergovernmental nature as well as by their varying relationship with the private sector.

REFERENCES

AARON, H.J. (1969). "Local public expenditures and the 'migration effect.' " Western Economic Journal, 7(4):385-390.

ADAMS, A.V., and NESTEL, G. (1976). "Interregional migration, education, and poverty in the urban ghetto: Another look at black-white earnings differentials." Review of Economics and Statistics, 58(2):156-166.

AIGNER, E.J., and HEINS, A.J. (1967). "On the determinants of income inequality." American Economic Review, 57(1):59-72.

AL-SAMARRIE, A., and MILLER, H.P. (1967). "State differentials in income concentration." American Economic Review, 57(1):59-72.

BAHL, R.W. (1968). "Studies on determinants of public expenditures: A review." Pp. 184-207 in J.J. Muskin and J.F. Cotton, Functional federalism: Grants-in-aid and PPB systems. Washington, D.C.: State-Local Finances Project of the George Washington University.

——— (1969). Metropolitan city expenditures: A comparative analysis. Lexington: University of Kentucky Press.

BEATON, W.P. (1974). "The determinants of police protection expenditures." National Tax Journal, 27(2):335-349.

BERRY, B.J.L. (1963). "Commercial structure and commercial blight." Chicago: Department of Geography Research Paper No. 85.

——— (1973). Growth centers in the American urban system. Cambridge: Ballinger.

BOOMS, B.H., and HU, T. (1973). "Economic and social factors in the provision of urban public education." American Journal of Economics and Sociology, 32(1):35-43.

BRADFORD, D.F., and KELEJIAN, H.H. (1975). "An econometric model of the flight to the suburbs." Journal of Political Economy, 83(4):566-589.

CASPARIS, J. (1967). "Metropolitan retail structure and its relation to population." Land Economics, 43(2):212-218.

CHAPMAN, J.L. (1976). "The demand for police." Public Finance Quarterly, 4(2):187-206.

CONROY, M.E. (1975). The challenge of urban economic development. Lexington, Mass.: D.C. Heath.

CROWLEY, R.W. (1970). "An empirical investigation of some local costs of inmigration to cities." Journal of Human Resources, 5(1):11-23.

DANZIGER, S. (1976). "Determinants of the level and distribution of family income in metropolitan areas, 1969." Land Economics, 52(4):467-478.

FARBMAN, M. (1973). "Income concentration in the southern United States." Review of Economics and Statistics, 55(3):333-340.

--- (1975). "The size distribution of family income in U.S. SMSAs, 1959." Review of Income and Wealth, 21(2):217-237.

FARLEY, R. (1977). "Trends in racial inequalities: Have the gains of the 1960s disappeared in the 1970s?" American Sociological Review, 42(2):189-208.

FORD, E.J. (1977). "Explaining interurban variation in the level and distribution of income." Review of Social Economy, 35(1):67-77.

FRIEDMAN, J.J. (1973). "Variations in the level of central business district retail activity among large U.S. cities: 1954 and 1967." Land Economics, 49(3):326-335.

FRISBIE, W.P., and NEIDERT, L. (1977). "Inequality and the relative size of minority populations: A comparative analysis." American Journal of Sociology, 82(5):1007-1030.

GOETTEL, R.J., and FIRESTINE, R.E. (1975). "Declining enrollments and state aid: Another equity and efficiency problem." Journal of Education Finance, 1(2):205-215.

GREENE, K.V. (1977). "Spillovers, migration and public school expenditures: The repetition of an experiment." Public Choice, 29:85-93.

GREENWOOD, M.J. (1976). "A simultaneous-equations model of white and non-white migration and urban change." Economic Inquiry, 41(1):1-15.

GUNTHER, W.D., and LEATHERS, C.G. (1975). "Trends in income inequality in the south, 1950 to 1970." Growth and Change, 6(2):19-22.

GUSTMAN, A.L., and PIDOT, G.B., Jr. (1973). "Interactions between educational spending and student enrollment." Journal of Human Resources, 8(1):3-23.

HAMILTON, J.R. and REID, R. (1977). "Diseconomies of small size and costs of migration." Growth and Change, 8(1):39-44.

HARRISON, B. (1974). Urban economic development: Suburbanization, minority opportunity, and the condition of the central city. Washington, D.C.: Urban Institute.

HERMALIN, A.I., and FARLEY, R. (1973). "The potential for residential integration in cities and suburbs: Implications for the busing controversy." American Sociological Review, 38(5):595-610.

JONISH, J.E., and KAU, J.B. (1973). "State differentials in income inequality." Review of Social Economy, 31(2):179-190.

KELLEY, A.C. (1976). "Demographic change and the size of the government sector." Southern Economic Journal, 43(2):1056-1066.

KUZNETS, S. (1955). "Economic growth and income inequality." American Economic Review, 45(1):1-28.

LONG, J.E., RASMUSSEN, D.W., and HAWORTH, C.T. (1977). "Income inequality and city size." Review of Economics and Statistics, 59(2):244-246.

LONG, L.H. (1974). "Poverty status and receipt of welfare among migrants and nonmigrants in large cities." American Sociological Review, 39(1):46-56.

--- (1975). "How racial composition in cities changes." Land Economics, 51(3):758-767.

LONG, L.H., and HANSEN, K. (1977). "Selectivity of black return migration to the south." Paper presented at the annual meeting of the Southern Sociological Society, April 1977.

MANNERS, G. (1974). "The office in metropolis: An opportunity for shaping metropolitan America." Economic Geography, 50(2):93-110.

MATTILA, J.M., and THOMPSON, W.R. (1968). "Toward an econometric model of urban economic development." Pp. 63-80 in H.S. Perloff and L. Wingo, Jr. (eds.), Issues in urban economics. Baltimore: Johns Hopkins University Press.

McDONALD, J.F. (1975). "Some causes of the decline of central business district retail sales in Detroit." Urban Studies, 12(2):229-233.

PAGLIN, M. (1975). "The measurement and trend of inequality: A basic revision." American Economic Review, 65(4):598-609.

PERLOFF, H.S., DUNN, E.S., Jr., LAMPARD, E.E., and MUTH, R.F. (1960). Regions, resources, and economic growth. Baltimore: Johns Hopkins University Press.

PERSKY, J.J., and KAIN, J.F. (1970). "Migration, employment, and race in the deep south." Southern Economic Journal, 36(3):268-276.

RAYMOND, R. (1972). Determinants of nonwhite migration during the 1950s: Their regional significance and longterm implications." American Journal of Economics and Sociology, 31(1):9-20.

REES, J. (forthcoming). "Manufacturing change, internal control and government spending in a growth region of the United States." In F.E.I. Hamilton (ed.), Industrial movement and change: International experience and public policy. London: Longmans.

RICE, G.R., and SALE, T.S., III (1975). "Size distribution of income in Louisiana and other southern states." Growth and Change, 6(3):25-33.

ROOF, W.C., VAN VALEY, T.L., and SPAIN, D. (1976). "Residential segregation in southern cities: 1970." Social Forces, 55(1):59-71.

RUTHENBERG, D., and STANO, M. (1977). "The determinants of interstate variations in income distribution." Review of Social Economy, 35(1):55-66.

SALE, T.S. III (1974). "Interstate analysis of the size distribution of family income, 1950-1970." Southern Economic Journal 40(3):434-441.

SHIN, E.H. (1975). "A cohort analysis of black-white income differentials in the south, 1960-1970." Review of Regional Studies, 5(3):14-30.

SOLTOW, K. (1960). "The distribution of income related to changes in the distribution of education, age, and occupation." Review of Economics and Statistics, 42(4):450-453.

STRUYK, R.J. (1972). "Evidence on the locational activity of manufacturing industries in metropolitan areas." Land Economics, 48(4):377-382.

TAUSSIG, M.K. (1977). "Trends in inequality of well-offness in the United States since World War II." Pp. 1-82, Part I of a Conference on the Trend in Income Inequality in the U.S., Institute for Research on Poverty, University of Wisconsin-Madison, Madison, Wisconsin, October 29-30, 1976.

THOMPSON, W.R. (1975). "Economic processes and employment problems in declining metropolitan areas." Pp. 187-196 in G. Sternlieb and J.W. Hughes (eds.), Post-industrial America: Metropolitan decline and inter-regional job shifts. New Brunswick, N.J.: Rutgers University Center for Urban Policy Research.

VAN VALEY, T.L., ROOF, W.C., and WILCOX, J.E. (1977). "Trends in residential segregation: 1960-1970." American Journal of Sociology, 82(4):826-844.

WEINSTEIN, B.L., and FIRESTINE, R.E. (forthcoming). Regional growth and decline in the United States: The rise of the sunbelt and the decline of the industrial northeast. New York: Praeger.

WEISS, L., and WILLIAMSON, J.G. (1972). "Black education, earnings, and interregional migration: Some new evidence." American Economic Review, 62(3):372-383.

9

Multinational Corporations, International Finance, and the Sunbelt

ROBERT B. COHEN

□ THE RAPID GROWTH OF THE SUNBELT STATES during the past two decades has led several writers to conclude that there has been a definite shift of economic power from the older urban centers in the Northeast and Central states to the expanding metropolises of the South and Southwest. In his well-known book, *Power Shift,* Kirkpatrick Sale sets the tenor of the argument, asserting that the growth of new technology and the increased private sector reliance upon government support resulted in "an authentic economic revolution that created the giant new postwar industries of defense, aerospace, technology, electronics, agribusiness, and oil-and-gas extraction, all of which were based primarily in the Southern Rim and which grew to rival and in some cases surpass the older industries of the Northeast" (Sale, 1976:6). This analysis is often combined with the assertion that the Sunbelt is now the decisive force in presidential and legislative politics. Sale's vision of the primacy of the Sunbelt in Congress, Nixon's "Southern Strategy" (Phillips, 1969) and the Carter presidency all serve to reinforce this line of reasoning.

AUTHOR'S NOTE: *I would like to thank Lynn Holland for her help in collecting and preparing the data cited in this paper.*

This argument, found both in domestic social science literature and the popular press, offers as prime evidence of a power shift a series of important economic trends which have preferentially favored the growth of the Southern Rim. The first and most important of these is the massive population shift. The second is the nation's shift away from an economy centered on the traditional heavy manufacturing sectors in which the industrial Northeast specialized to the new technology and energy industries, which were most frequently located in the Sunbelt. Finally, various studies have argued that the replacement of blue-collar jobs with service and government occupations has been especially beneficial to the Sunbelt.

In all, these interregional trends have, in Sale's terms, given rise to a "rival nexus" in which the Sunbelt, "moving on to the national stage and mounting a head-on challenge to the traditional Establishment has quite simply shifted the balance of power in America away from the Northeast and toward the Southern Rim" (Sale, 1976:6).

However, as this paper will demonstrate, most corporations in the South and a large number of those in the Southwest continue to depend upon non-Sunbelt banks to finance their international and domestic operations. These corporations also rely heavily upon large law firms outside the Sunbelt and upon the major accounting firms, all of which are headquartered in the Northeast and North Central states. As a result, a number of variables can be identified which suggest that the economic rise of the Sunbelt does not portend a shift in the traditional centers of corporate and financial power. More specifically, it is hypothesized that despite the widely heralded regional realignment of population and industrial activity, there is little evidence to suggest that this has been accompanied by a decisive shift in the traditional locations of corporate control functions. Rather, a more apt analogy is an athletic team which now plays its games on a new and expanded field but does not change its coaches and other directional personnel.

In conjunction with this argument, power will be defined as the corporate and financial control of an economy. This paper will then examine whether Sale's analysis of a power shift is corroborated by various indices of the location of corporate control functions. I shall argue that: (1) non-Sunbelt firms continue to dominate the leading control sectors of the economy, although Sale claims Sunbelt firms dominate these activities; (2) non-Sunbelt firms control a large share of the ongoing investment in the Sunbelt; and (3) Sunbelt corporations are closely tied to the traditional centers of the Northeast and North Central regions through their banking, legal, and accounting relationships. Thus, the essential elements of a fundamental realignment in the centers of corporate power are not supported by this preliminary data.

The first section discusses the nature of the modern corporation and its relationship to the growth of the Sunbelt, with particular emphasis on the international expansion of U.S. corporations and their research and development activities. Section two examines the linkages between Sunbelt corporations and important corporate service firms, including banks, law firms, and accounting firms. The concluding section summarizes how changes in the modern corporation have altered the economic role of the Sunbelt and suggests that traditional analytical frameworks must begin to incorporate an understanding of corporate dynamics and hierarchy if they wish to explain regional change.

THE CORPORATION AND THE GROWTH OF THE SUNBELT

Most analyses of a "power shift" (Sale, 1976) conclude that since industrial expansion in the Sunbelt has been more rapid than in other regions, a power shift has taken place. However, despite these undeniable statistics, if the growth is fueled primarily by outside investors, then the pace of development is not subject to internal control. Under these circumstances, the notion of a "power shift" becomes more tenuous.

As suggested above, Sale posits that power in the Southern Rim is based upon rapidly growing new industries, key links to the federal government, and a series of shifts of traditional control centers to the Sunbelt. Implicit in his argument is the assertion that firms in the Southern Rim are an important element in this transformation. However, this analysis is seriously deficient in several respects. First, the evidence of a shift in power is based primarily on aggregate industrial data. However, the control functions which are the major repositories of economic power are never identified nor are the location proclivities of these crucial activities specified in any detail. In addition, the ownership of "sunbelt" corporations is scarcely discussed. Second, since most studies focus upon regional industrial information, they emphasize the rise of industries which are oriented to new technologies, but fail to examine broader systemic changes in the economy, such as the rapid postwar expansion of foreign investment by U.S. corporations. Finally, Sale argues that even if New York financiers own a mine located in Arizona, "a greater percentage of the profits, and virtually all other money, will stay right there in Arizona" (Sale, 1976:65). This may be true in some cases, but outside ownership is a fragile guarantee of continued reinvestment in the local economy.

In addition, although regional analysts note the rapid growth of cities in the Southern Rim, they do little to identify the corporate actors who are sponsoring this growth. There is a general recognition, for example, that Houston is the capital of the Sunbelt and that it is also the center of the petroleum industry and of the national space program. However, they have thus far failed to examine the corporations which have aided the growth of other Southern Rim cities and have yet to adequately compare the corporate builders of Houston with the corporate giants of other "super cities," such as New York, Chicago, Los Angeles, and San Francisco. Consequently, we are provided with an argument which is in some ways sound, but which is incomplete for inferring a regional "shift in power," defined as a shift in the locus of corporate and financial control of the economy.

THE MODERN CORPORATION AND URBAN DEVELOPMENT

In recent years several economists have become increasingly concerned with the role of corporations in urban development. Hymer, for instance, has noted that the modern corporation is composed of three functional levels: level three, the lowest, is concerned with day-to-day operations; level two, which emerged with the separation of headquarter functions and field offices, coordinates management at level three; and level one includes top management for goal-determination and planning. The spatial separation of these functions parallels the spatial expansion of corporations in their drive towards national and multi-national status (Hymer, 1972:123-124).

Hymer has also argued that the three levels of decision making correspond to the spatial distribution of corporate functions and power. Thus, he suggests that level one activities are concentrated in world cities, level two activities are largely located in national cities, and level three activities are mostly in regional cities or else they are dispersed on a worldwide basis.

Hymer's level one cities are characterized by concentrations of media, government, and capital markets and supporting services which interface with these activities, such as lawyers, business services, and top level professionals; consultants, designers, artists, specialists, large wholesalers and brokers (Hymer, forthcoming). My own research (Cohen, forthcoming) has also focused upon the corporation and several of the important services which Hymer believed would be concentrated in level one cities. I have discovered that with the emergence of the modern corporation, several sophisticated business services were transformed from a supporting resource for major corporations to key inputs into corporate decision making. The structure of the accounting, legal, banking, and investment banking professions changed substantially, resulting in a division of labor whereby only a few firms offered the most sophisticated skills and strategic information. These firms are no longer merely "service" corporations; instead they represent critical inputs for formulating corporate strategy. Hence, the largest and most important banks, law firms, and accounting firms are largely concentrated in only a few world cities, for example, New York and San Francisco and several national cities such as Chicago and Los Angeles.

Therefore, it is possible to construct an urban hierarchy in the United States by analyzing the urban location of various corporate functions, among which are foreign operations of corporations; corporate links with law firms, banks, and accounting firms; and the range of functions provided to corporations by sophisticated service firms. In addition, an examination of the complementary hierarchies of banks, law firms, and accounting firms reveals that they are even more concentrated than corporate headquarters. Hence with the entrance of many large U.S. corporations into the world market during the postwar period, it has become even more important for businesses to have well-established links with the critical service firms. Long-range strategy formulation has become a major concern of the modern corporation, formal intracorporate linkages have been established which avoid market transactions, and greater emphasis has been placed on the control of financial and production information and the need to control a global labor force.

Thus, if we define power in terms of corporate control of economic affairs, we can supply a much needed link in the chain of analysis concerning the economic-political relationship between the industrial Northeast and the rising Sunbelt. It is particularly important in this respect to investigate whether Sunbelt firms have been major participants in the overseas expansion of corporate activity since this is perhaps the most dynamic and profitable corporate function in the postwar period. In addition, by using corporate data for research and development expenditures and illegal payments abroad, it should be possible to determine whether the two industrial activities which analysts claim are the exclusive province of Southern Rim firms–technological sophistication and political influence–are truly concentrated in this region. Also, statistics displaying both the incidence and location of foreign investment in U.S. industry and the role of non-Sunbelt banks in Sunbelt financial affairs should allow us to develop rough indices of the depth of external control over the economy of the Sunbelt. Finally, focusing on corporations permits us to study the ties between corporations in the Sunbelt and banks, law firms, and accounting firms located throughout the world. This too should provide insight into the relative dependence or autonomy of Sunbelt corporations and, hence, of their economic power.

FOREIGN SALES OF MULTINATIONAL CORPORATIONS

Perhaps the most startling change in corporate activities during the postwar period has been the growth of foreign subsidiaries. We can use total foreign sales to measure these operations and then compare these figures with total corporate sales. Since the international expansion of U.S. firms has been such an important ingredient in the transformation of the nation's economy, this should enable us to determine if Sunbelt firms are key participants in these developments.

The most important urban centers of international corporate activity are listed in Table 1. When we calculate the ratio of foreign sales to total sales for those firms whose headquarters are located in each city, a number of cities emerge as prominent international business centers. Corporations in New York, San Francisco, Pittsburgh, and Houston have foreign sales which are a third or more of

TABLE 1
MAJOR CENTERS OF INTERNATIONAL BUSINESS IN THE UNITED STATES

SMSA	Foreign Sales[a] of *Fortune 500 Firms* Headquartered in SMSA-1974 ($ billions)	Share of Total Foreign Sales (%)	Total Sales of *Fortune 500 Firms* Headquartered in SMSA-1974 ($ billions)	Share of Total Sales (%)
New York	98.9	40.5	252.9	30.3
Detroit	21.5	8.8	76.0	9.1
Pittsburgh	15.2	6.2	41.0	4.9
San Francisco	13.1	5.4	26.5	3.2
Chicago	11.2	4.6	60.9	7.3
Los Angeles	9.3	3.8	38.4	4.6
Houston	6.7	2.8	17.4	2.1
Akron	4.2	1.7	12.6	1.5
Minneapolis	3.1	1.3	13.4	1.6
Patterson	2.9	1.2	10.1	1.2
St. Louis	2.9	1.2	17.3	2.1
Cleveland	2.7	1.1	17.2	2.1
Top 50 SMSA's	212.5	87.0	691.7	82.9

a. Includes only the foreign sales of firms reporting them, or 255 of 386 corporations in the top 50 SMSAs.
SOURCES: *Fortune,* May 1975, and a survey of the foreign sales of U.S. corporations prepared from 10-K Reports and prospectuses filed by corporations with the S.E.C. (Cohen, forthcoming).

their total sales. On the other hand, some cities have a number of corporations which are not as active overseas as they are domestically. These places have firms whose ratio of foreign sales to total sales is .25 or less; i.e. Los Angeles, Chicago, St. Louis, and Cleveland.

From these results, corporations in San Francisco and Houston do appear to be as involved overseas as their counterparts in New York and Pittsburgh, and more active than firms headquartered in other traditional Northeastern and North Central centers. It should be noted, however, that most of the cities in this latter group are most heavily tied to regional economies, and hence their firms are more oriented to domestic, rather than international, pursuits. With the exception of Chicago, these cities are not centers of aggressive international banking activity, international financial, and strategic business services.

While San Francisco and Houston appear to be important international centers, they are relatively weak when compared with New York. One explanatory factor is the dependence of Sunbelt corporations, particularly those in Houston, upon the international banking services they obtain from banks outside the Sunbelt, mainly in New York (Stodden, 1973). In addition, many of the international accounting and law firms employed by these corporations are located in New York. This has important implications for the development of these two Sunbelt centers as world cities since it implies that New York offers many more locational advantages to firms which want international business links. Thus, the absence of these strategic services may prove detrimental to their ability to compete with the international expansion of their Northeastern counterparts (Cohen, forthcoming).

If the Sunbelt is not the center for many multinational corporations, it does have several cities which are notable regional centers. Coral Gables, outside Miami, has become the nation's "Gateway City to Latin America," housing the Latin American regional offices of 55 U.S. multinationals and two British multinationals. The Sunbelt/South now has regional offices for 19% of the *Fortune* 500 corporations and the Sunbelt/West houses an additional 16% of these regional offices (Bruner, 1976). Thus, although the Sunbelt cities are not the home for most of the major multinationals, they are emerging as important regional centers. But despite these developments, the major control center for U.S. international business still remains the old and traditional Yankee capital, New York City.

RESEARCH AND DEVELOPMENT EXPENDITURES

Citing the sprawling scientific complexes of Los Angeles, Phoenix, Dallas, and other Southern Rim cities and the rapid expansion of professional and technical employment, Sale asserts that the area "has been the main beneficiary of the growth of the technology industry" (Sale, 1976:31). However, a detailed scrutiny of the data casts some doubt on this assertion.

Recent changes in corporate accounting practices have standardized the reporting of research and development expenditures. I have utilized the first published survey of R & D spending and assigned these expenditures to the region housing the corporate headquarters on the assumption that research facilities are generally located in close proximity to the head office. The current data excludes government support of private sector R & D, and while it therefore ignores an important component of the total effort, it does provide a clear indication of the firm's commitment to innovation.

Table 2 compares R & D expenditures for Sunbelt and non-Sunbelt corporations. Surprisingly, only 6.46% of the private R & D efforts is attributable to firms headquartered in the Sunbelt, while 93.53% is generated by non-Sunbelt corporations. Nearly 40% of R & D spending is reported by firms headquartered in New York and Detroit, while Los Angeles's corporations account for 2.46% of all such spending and firms in San Francisco and San Jose have but 0.83% and 0.85%, respectively.

One explanation for this disparity is that the older research-oriented corporations of the Northeast and Central states have been forced by foreign competitors to diversify and to develop a great number of new products. This forces them to maintain a high level of expenditures. In the Sunbelt, however, the heavy infusions of government support have blunted the private sectors' commitment to finance these costs from their own treasury. An initial examination of government R & D contracting with Sunbelt corporations further substantiates this pattern of governmental subsidization in the Southern Rim and private initiative in the Northeast (U.S. Department of Defense, 1976). Thus, in R & D terms, we are overestimating the role of the Southern Rim as a center for the development of U.S. technology. Significantly more money still is spent on research outside the Sunbelt.

TABLE 2
R & D SPENDING BY FORTUNE 500 CORPORATIONS IN 1975

	Fortune 500 Corporation Reporting R & D Expenditures	Total R & D Expenditures[a] ($ millions)	Share of Total R & D Expenditures (%)
Sunbelt corporations (92)[b]	52	822.6	6.46
Top 50 SMSAs (72)	42	742.7	5.83
Outside top 50 SMSAs (20)	10	79.9	.63
Non-Sunbelt corporations (408)	268	11,909.0	93.54
Top 50 SMSAs (314)	201	9550.6	75.02
Outside top 50 SMSAs (94)	67	2358.4	18.52
Total corporations (500)	320	12,731.6	100.00

a. Only corporations whose R & D expenditures exceed one percent of total sales are required to report such spending.
b. Total number of Fortune 500 corporations in each category is given in parentheses.
SOURCE: *Business Week,* 1976. Only corporations on the Fortune 500 list were used.

LINKAGES BETWEEN CORPORATIONS AND CORPORATE SERVICES

BANKING AND INTERNATIONAL FINANCE IN THE SUNBELT

The Foreign Expansion of U.S. Banks. According to Sale, one of the major factors contributing to the regional power shift is the ability of Southern Rim corporations to obtain funds from the federal treasury. In addition, he posits that the rapid growth in this region is fueled by more adventuresome entrepreneurs who are not afraid to take large financial risks. If this were true, one might expect to find banks in the Sunbelt providing more funds to the overseas operations of U.S. firms and financing most of the region's capital requirements.

One way to examine the links between U.S. firms abroad and Sunbelt banks is to compare the ratio of foreign deposits in these banks to their domestic deposits. This provides a surrogate measure for the links between Sunbelt banks and the overseas operations of U.S. corporations. Such data indicate that banks in the Sunbelt are relatively less active overseas than banks elsewhere in the U.S. (Table 3). The data on foreign bank deposits reveal that New York is clearly the most important international financial center with nearly 50% of its deposits originating abroad. The two most active international banking centers in the Sunbelt are San Francisco and Dallas with 33% and 25%, respectively, of their total deposits originating from nondomestic sources.

TABLE 3
DOMESTIC AND FOREIGN DEPOSITS OF THE TOP 200 U.S. BANKS IN 1976

	Total Domestic Deposits ($ millions)	Share of Top 200 Banks (%)	Total Foreign Deposits ($ millions)	Share of Top 200 Banks (%)
Sunbelt banks (63)[a]	162,987	34.60	38,246	23.88
Top 50 SMSAs (42)	137,274	29.15[b]	37,273	23.27
Outside top 50 SMSAs (21)	25,713	5.46	973	.61
Non-Sunbelt banks (137)	308,024	65.40	121,947	76.12
Top 50 SMSAs (113)	283,052	60.10	121,831	76.05
Outside top 50 SMSAs (24)	24,972	5.30	116	.07
Total (200)	471,011	100.00	160,193	100.00

a. The number of banks in each area is given in parentheses.
b. Error is due to rounding.
SOURCE: *Business Week*, 1977. (Data: Investors Management Sciences, Inc.)

In international banking, size is all-important, since only the largest banks can provide the services needed by major corporations. Foreign trade financing requires that a bank have a foreign exchange arbitrageur, but only the top 20 banks offer this service. These 20 banks can take advantage of the federal funds market much more easily than their smaller competitors (Muller and Cohen, 1977). In San Francisco, where several banks are among the top 20, the recent foreign expansion has been particularly explosive. Between 1969 and 1974, total foreign assets of West Coast banks increased from $7 billion to $32 billion (Cheng, 1976:10-11). U.S. capital outflows reported by these banks averaged $3.2 billion per year, or 18 times the average rate during the 1969-1970 period. Cheng notes that the recent growth of many West Coast banks during the early 1970s may have been a crucial factor in their decision to enter the international finance arena. In addition, since several large and internationally active banks are already located in San Francisco, local competitors may have been forced to offer these services in order to retain their domestic business.

Yet while West Coast banks are becoming more involved internationally, those in other parts of the Sunbelt, particularly the Sunbelt South, have had great difficulty establishing themselves on the international scene. There have been a number of recent changes which may alter this situation, but many of the problems facing these banks seem almost insurmountable. In international finance, a late start is a disadvantage which is almost impossible to overcome.

Bank Links With Corporations. Throughout the Sunbelt South, the lack of sufficient bank capital has forced the major financial institutions to forgo lucrative financial arrangements with many of the region's leading corporations. In Texas, for instance, more than half of the 42 largest corporations with over $18 billion in assets, reported that their company's principal bank was out of state (Stodden, 1973:6). Most named a bank in New York as their principal bank. Nearly three-fourths of these firms' out-of-state banking ties were with banks in New York, Chicago, and California. In 1972 alone, more than $500 million in loans and deposits left the state (Stodden, 1973:6-7).

In roughly three-fifths of the cases where Texas banks are used by Texas corporations, these banks must join with large out-of-state banks to satisfy the financing requirements. Even though the emergence of larger holding company banks may ameliorate this

TABLE 4

COMMERCIAL AND INDUSTRIAL LOANS OUTSTANDING, SELECTED FEDERAL RESERVE DISTRICTS

Business of Borrower	Total Loans All Districts ($ millions)	Loans by New York Banks % of U.S. Total	Loans by Atlanta Banks % of U.S. Total	Loans by Dallas Banks % of U.S. Total	Loans by San Francisco Banks % of U.S. Total
Durable goods manufacturing	14,877	34.15	2.61	3.05	12.46
Nondurable goods manufacturing	13,816	47.12	4.92	3.25	9.16
Textile apparel and leather	3,375	47.38	9.21	1.60	4.15
Petroleum refining	2,348	54.98	1.19	9.16	4.86
Mining (including petroleum and natural gas)	7,436	40.18	2.53	17.00	11.73
Trade	15,402	33.46	5.09	3.72	13.97
Construction	3,974	20.99	7.02	11.73	14.42
Bankers acceptances	3,903	33.54	.90	5.71	21.55
Foreign commercial and industrial loans	5,834	42.17	1.97	3.24	29.26
Total commercial and industrial loans[a]	116,774	33.00	4.15	5.05	20.04

a. The following major categories of commercial and industrial loans are not presented in this table but are included in the total: transportation, communication and other public utilities; services; and all other loans.

SOURCE: Federal Reserve Bank, 1977. The data are taken from a sample of large commercial banks which reported for the week ending March 30, 1977.

condition, Texas banks are still quite deficient in international banking services. This causes many companies to rely on New York banks (Stodden, 1973:6) and this is reflected in the data on commercial and industrial loans issued by banks in the Atlanta and Dallas Federal Reserve districts. These banks make substantially fewer loans to corporations than do their competitors in the New York and San Francisco districts (see Table 4).

Non-Sunbelt Banks in the Sunbelt. A further indication of the weakness of most Sunbelt banks in international financial affairs is apparent from the recent growth of Edge Act subsidiaries throughout the Sunbelt. Banks are permitted to establish these offices outside their home state if they will exclusively handle international finance. They are also often set up at a bank's head office. In recent years, banks in the Northeast have used these subsidiaries to take even more business away from banks in the Sunbelt. Between 1970 and 1975, these banks established numerous fully capitalized Edge Act subsidiaries throughout the Southern Rim (see Table 5), and as of 1975, the local bank's Edge Act offices in the Sunbelt accounted for only one-third of the total capital used to establish such offices in the Sunbelt.

In terms of international finance, only San Francisco seems to have emerged as a major center in the Sunbelt. While Dallas is an important financial center, recent work has shown that it is losing ground to New York (Ahlers, 1977). Thus, even though some Sunbelt firms are active abroad, it appears as though their reliance upon outside banks, except in the case of those in San Francisco, will continue for the foreseeable future.

BUSINESS SERVICE LINKAGES AND THE SUNBELT

Another means of analyzing the strength of the challenge Sunbelt industries pose to the traditional Establishment (Sale, 1976) is to examine linkages between Sunbelt corporations and financial, legal, and accounting firms.

Recent evidence suggests that the traditional ties between New York financial firms and corporations located throughout the nation are extremely strong. Also, links between New York law firms, accounting firms, and corporations located outside New York are also impressive. Among the corporations headquartered outside New York, nearly all use investment bankers in New York and most have

TABLE 5
EDGE ACT CAPITALIZATION BY REGION

	Total Edge Capitalization (thousands of dollars)		Edge Capitalization belonging to Banks in Region (thousands of dollars)		Percent Edge Act Assets "Internally capitalized"	
Regions	1970	1975	1970	1975	1970	1975
Southeast[a]	14,499	69,273	14,499	23,329	100	33.7
Texas[b]	4,238	41,784	4,238	16,296	100	39.0
Southwest[c]	41,336	172,293	23,689	56,294	57.3	32.7
New York[d]	466,924	981,740	213,599	491,677	45.7	50.1
Chicago[e]	64,724	133,733	64,724	110,648	100	82.7

a. Includes Winston-Salem, Richmond, Norfolk, Miami, and Atlanta.
b. Includes Dallas, Houston, and New Orleans.
c. Includes San Francisco and Los Angeles.
d. Includes only New York City.
e. Includes Detroit and Chicago.
SOURCES: *The Journal of Commerce,* 1971, 1976; Federal Reserve Bank of New York, Foreign Banking Department, 1976.

their major bank lender and pension fund administrator in New York. Over 350 of the *Fortune* 500 corporations use accounting firms based in New York. Among the nine corporations on the *Fortune* 500 list which are headquartered in Houston, four had their bank lenders in New York and three had pension fund administrators in New York. Only in the case of legal services do local firms receive the bulk of the business (Cohen, forthcoming).

These strong ties to New York are still in force because financial and accounting firms located in New York City (and law firms, to a more limited degree) have great advantages over their rivals. They have the best skills, the quickest access to information, and the most extensive contacts with major corporations in the U.S. and overseas. Such advantages are difficult to overcome, particularly if important services like banking and investment banking are underdeveloped in a region or city, as they are in Houston.

SUMMARY AND CONCLUSIONS

This discussion is not meant to obviate the impressive array of evidence that has been marshalled recently detailing the growth of the urban Sunbelt. The growth of new technological industries, an increasing set of benefits accruing from realized political power, and the critical developmental role of the federal government have all contributed to a revitalized and booming region. However, such trends should not be read as a total picture of the configuration of American corporate, locational, and financial power. The corporations headquartered in the Southern Rim have not been particularly active overseas nor have they been major spenders on R & D. The linkages between banks in the Sunbelt and local corporations are weak, and the region's banking sector, with the exception of banks in San Francisco, does not compete with banks in Chicago and New York.

Is there some way to explain the rapid industrialization of the Southern Rim and the small role its corporations have played in such development? As the modern corporation diversified and invested abroad, it sought new markets and attractive conditions for production. Like many developing nations, the Southern Rim offered both. Numerous factories moved South and many foreign firms invested there. Thus from a corporate power standpoint, the rise of the Sunbelt, like the development of a third world nation, can be viewed as a new region's integration into the world as well as the national economy. And like a dependent nation, much of the Sunbelt's industry and a significant portion of its finances remain under the control of outside economic actors.

REFERENCES

AHLERS, D.M. (1977). "A new look at U.S. commercial banking." Executive, 3(2):8-22.
BRUNER, D. (1976). "A large number of corporations establishing offices in the south." American Banker, May 6, p. 10.
Business Week (1967). "Companies hark to Coral Gables' Spanish accent." April 22, pp. 123-126.
--- (1976). "R & D spending: 1975." June 28, pp. 63-84.
--- (1977). "Annual survey of bank performance." April 18, pp. 98-106.
CHENG, H.S. (1976). "U.S. west coast as an international financial center." Federal Reserve Bank of San Francisco Economic Review, (spring):9-19.
COHEN, R.B. (forthcoming). The modern corporation and the city.
Federal Reserve Bank (1977). "Statistical release." April 6.
Federal Reserve Bank of New York, Foreign Banking Department (1976). "Total equity capital accounts of corporate organization organized under section 25 (A) as of 12/31/75." Unpublished data.
Fortune (1975). "The Fortune 500." May.
HYMER, S.H. (1972). "The multinatinal corporation and the law of uneven development." Pp. 113-140 in J.W. Bhagwati (ed.), Economics and the world order. New York: Macmillan.
--- (forthcoming). "The multinational corporation and the international division of labor." In S.H. Hymer, The multinational corporation: A radical critique (R. Cohen et al., eds.). Cambridge: Cambridge University Press.

Journal of Commerce (1971). "Directory of Edge Act affiliates of U.S. banks." December 13, pp. 24, 25, 27, 31.

--- (1976). "Directory of Edge Act affiliates of U.S. banks." December 13, pp. 8-10.

MULLER, R.E., and COHEN, R.B. (1977). "The transformation of U.S. banking: A systemic dilemma." Executive, 3(2):37-44.

PHILLIPS, K.P. (1969). The emerging republican majority. New Rochelle, N.Y.: Arlington House.

SALE, K. (1976). Power shift. New York: Vintage Books.

STODDEN, J.R. (1973). "Their small size costs banks business of large companies." Federal Reserve Bank of Dallas Business Review, October:6-7.

U.S. Department of Defense (1976). "The 500 largest military prime contractors for research, development, test and evaluation work." Washington, D.C.

Part III

Visions of the Urban Future

Introduction

□ IN THE PREVIOUS TWO SECTIONS, the authors analyzed some of the economic forces promoting the rise of the Sunbelt cities as well as some of the social and political dynamics which are characteristic of this region. At this point it seems appropriate to step back from the details and reassess some of the major implications associated with this transformation of the economic landscape.

The three selections comprising this final section are broad, wide ranging, discursive essays which address a number of major issues that the nation will have to grapple with in the coming decades. If there is one common theme in all three pieces it is that complacency in the Sunbelt and unmitigated pessimism in the Northeast are unwarranted. The general economic and demographic trends which have garnered so much attention are not immutably fixed and, thus, neither region can afford to march into the future with their prevailing business-as-usual attitudes intact. Rather, both sections will be forced to adjust if they are to continue to thrive and jointly contribute to the economic and social well-being of the nation.

But what sort of adjustments are necessary? The authors all agree that the future will require a redefinition of the nature of the city and its economic functions as they relate to the social and communal aspects of urban civilization. The future will require a reconciliation between the often contradictory goals of growth, the preservation of free markets, and social welfare. The balance between these goals and the weight given to each by the American polity will determine the future course of urban development in the United States.

Gurney Breckenfeld's prescription for the future is premised on the twin assumptions that it is possible for cities to grow old and even shrink gracefully and that the first incipient signs of ungainly aging are appearing in the Sunbelt. Thus, the redefinitional task for the Sunbelt is preventative; on the whole, the cities of this region have been proceeding correctly but, in some areas, they are falling prey to the tried and failing nostrums adopted in the Northeast. Unless the Sunbelt cities limit their sins of commission, their future will be as bleak as the Northeast's present. In the Northeast, however, the problem is more severe. According to Breckenfeld, their economic problems have been compounded by well-intentioned but misguided efforts at social engineering. These have created a perverse set of incentives which enervate their economic base and speed the cycle of decline, abandonment, deterioration, unemployment, and social disruption. Unless the political leaders in these cities are willing to bite the bullet and adopt new policies which improve the business climate of their locales, Breckenfeld feels that their decline will continue to be rapid and painful.

Breckenfeld assumes that the main thrust of our redefinitional policies must be focused in the economic arena. Unless the cities adopt programs which are more amenable to continued capital accumulation, they can expect little relief from their present depressed condition. Implicit in his analysis is the notion of an asymmetrical dependency between the city and the business community. The city is wedded to particular corporate entities for economic prosperity, but the corporation only views the city as another factor of production. Whenever the corporation perceives that the city can no longer serve its needs, whether because of a "bad business climate," an unfavorable political administration, or continued social disruption, the city will be discarded and replaced with a more pliant substitute. The next two selections accept this scenario but with one important caveat—its veracity depends on the extent of our devotion to the present social, economic, and institutional arrangements which dominate life in the United States. Once we begin to think in terms of altering these arrangements—which are all too frequently accepted as a priori limitations on the scope of social policy—then it is possible to redefine the nature of the city and our concept of success.

Specifically, this redefinition would start with the premise that despite its undeniable economic success, the rise of the Sunbelt represents continued human decline. Furthermore, if we consider people struggling to capture the city for themselves and for their own social betterment as a positive achievement, then the present situation in the Northeast is not necessarily indicative of "failing" cities. Quite the contrary, it is a reaffirmation of their vitality and of an ongoing positive struggle to convert the city from a place where people only work and struggle to survive to a place where people can live with respect and dignity.

Murray Bookchin argues that the Northeast is presently in the midst of this latter redefinitional struggle. At present, the major Northeastern metropolitan areas are being tugged in two diametrically opposed directions; while one group wants to redefine the city in terms of the account books where profits are the only relevant consideration, another group is seeking to return to the city of the history books where social interaction and human welfare reign supreme. If the latter group is victorious, then Bookchin would consider the cities in the Northeast to have fulfilled their primary function. They would be successes, not failures. In the Sunbelt, however, this struggle is less developed. The forces which are seeking to shape the destiny of these cities in terms of the account books are emerging victorious and thus, for Bookchin, the cities of the Sunbelt represent urban failures whose path should be shunned, not emulated.

David C. Perry and Alfred J. Watkins build upon Bookchin's analysis and extend it in several directions. In their view, the use of the term "the city" is a misnomer which fails to identify the contradictory layers which comprise urban civilization. In reality, "the city" is composed of two segments—an economic unit of production layer coupled with a social community strata. Although every city contains both components, the relative strength of each determines whether or not "the city" will survive in the capitalist milieu. Specifically, their arguments can be framed in terms of the following question: Why is poverty in the Sunbelt indicative of a good business climate while the same incidence of social distress in the Northeast is considered to be a symptom of decay? They respond by noting that poverty was always present in the Northeastern cities and that high unemployment, tenement housing, and inadequate incomes were never barriers to the city's previous growth as a unit of production. Only when these conditions provoke unrest in the social community and disturb the quiescent political environment does poverty assume the status of a barrier to economic growth. They go on to suggest that complacent poverty promotes growth; contentious poverty hastens the corporate exodus. By analyzing the relative strength of these strata in the Northeast and the Sunbelt, Perry and Watkins attempt to explore the shifting rhythms and patterns of uneven development as well as the basis by which American society defines urban success and failure.

10

Refilling the Metropolitan Doughnut

GURNEY BRECKENFELD

□ AMERICANS HAVE ALWAYS DISPLAYED a fascinating ambivalence about their cities. Thomas Jefferson regarded them as sinkholes of iniquity and corruption and a degree of anticity sentiment has persisted throughout U.S. history. Yet in the 19th century and the first part of the 20th, big cities led the nation's surge to industrialization and rising prosperity and fostered the great expansion of the middle class. Today, we commonly insist that we want our cities to be thriving centers of industry and commerce, of culture and amenity. Yet collectively we behave as though we do not mean it. For decades now, middle-class Americans have been demonstrating, by moving away from central cities, that they no longer choose to live in them, or even work in them if that can be avoided. Industry, office jobs, and retailing have joined the exodus, sapping the economic underpinnings of dozens of U.S. cities.

Over the past quarter century, hundreds of billions of dollars of public and private funds have been spent in efforts to "save" cities. Most of the programs have been misdirected and ineffectual and most of the money (especially public money) has been frittered away on the treatment of symptoms rather than the real causes of urban decline and decay. By avoiding difficult and politically unpopular decisions to take really meaningful action, we have unconsciously chosen to let a lot of our big cities rot. Increasingly that is what is happening to them. Many inner cities have become garbage dumps for the poor, the misfit, and the unwanted. The combination of crime, poor schools, high taxes, and high costs have drained jobs and population, vitality and spirit out of many cities to such an extent as to call into question their future economic and social role.

On their present course, quite a few big U.S. cities seem destined to become economic holes in the metropolitan doughnut. Their economies are floundering and their municipal finances are severely strained. Several have been flirting with bankruptcy. So far, the leading candidates for that unwelcome "hole-in-the-doughnut" status can be found among the old and ailing industrial cities of the Northeast and Midwest. In Detroit and St. Louis, the cycle of decline and the abandonment of housing, stores, and even factories has reached a point at which the two cities already can be considered as doughnut holes. Cleveland, Buffalo, and Newark, N.J., are borderline cases and many another Northeastern city is clearly sliding in the same direction. But even large Sunbelt cities, many of which have been rapidly gaining in population and economic strength in recent years, are by no means immune from such a development. Several of them, as we shall see later, already show early symptoms of ailments quite similar to those of cities in the Northeast and Midwest. (For this analysis, a city is "large" if it was among the top 50 in the census bureau's estimates of 1975 city populations. The 50th ranking city, Charlotte, N.C., had a 1975 population of 281,417.)

These necessarily subjective judgments are based on a quarter century of reporting, writing, and editing magazine articles and books about slums, housing, and other urban problems, and on a personal inspection of the ailing cities of the Northeast and Midwest late in 1976 and a similar journalistic foray into the Sunbelt in early 1977. I was, I thought, case hardened by years of exposure to slums not only in the U.S. and Europe but in such far-off places as Teheran and Hong Kong. Nonetheless, I found myslef shocked and depressed at the wreckage in American cities which I remembered from their livelier, healthier days 10 or 20 years ago.

Some of the world's big cities, notably in Europe, have grown old gracefully. Like many other American cities, Detroit and St. Louis have grown old disgracefully, though that does not mean that either should be considered as a terminal case. Their designation as doughnut holes means that they bear hideous scars from prolonged economic and population shrinkage and appear to have scant prospects for regaining their former industrial or commercial strength, or, in the near term, for reattracting many of the middle-class residents for which they yearn.

Though Detroit remains by a narrow margin the nation's fifth largest city (population: 1,335,085), in economic terms it may well be the sickest among the top 20. Its population has dropped about one-third from the 1959 peak of nearly 2,000,000. During the five years ending in 1976, the number of people at work in the city declined by a staggering 26%. Numerous factories built as recently as the 1940s and 1950s stand empty, for the city is long past its industrial prime and still going downhill. In much of the city the atmosphere is almost sepulchral, an eerie compound of emptiness and half-hidden menace. Downtown has become a hollow core pocked by vacant stores, boarded-up taverns, and the blank marquees of shuttered move theaters; the few theaters that remain show sex and violence films (e.g., "Snuff" and "Superfly") aimed at black audiences. An extraordinary number of tall office buildings seem about half empty, though still in good repair; one 23-story skyscraper was razed by its owners in 1976. Crowley's, the next-to-last department store downtown, closed in mid-1977. Surviving J.L. Hudons's, where sales have been declining for two decades, has blocked off large portions of its now excess space.

Nearby Washington Boulevard, formerly Detroit's equivalent of Manhattan's Fifth Avenue, is a lonely canyon of concrete and granite. Most of the shops that once offered the finest merchandise in town have closed or moved away. There, too, stands the empty and padlocked Statler Hotel, which was sold by the Hilton chain to another operator who failed. On a weekday midmorning, hardly a pedestrian is to be seen along the boulevard.

In the southwest corner of Detroit, several blue-collar ethnic neighborhoods (Italian, Polish, and German) remain in good order, their gardens tended, their alleys unlittered. Elsewhere, the city's extensive residential areas of aging one-family homes are laced with abandoned and vandalized stores and dwellings. Many of the houses were unwisely sold in the late 1960s under federally subsidized housing programs to families who could not afford them. The federal Department of Housing and Urban Development (which is, by a wide margin, the nation's biggest slumlord) had managed by late 1976 to get rid of almost half of the 13,700 foreclosed abandoned homes it had on its hands two years earlier, mainly by tearing the houses down and selling the lots to the city for $1 apiece. But Elmer Binford, the area HUD director, figured that there were still another 15,000 abandoned homes in private ownership throughout the city. The Detroit Free Press reported that half the people it queried in a poll "would move out of the city if they could afford to." Many could not, because they could only sell their homes at a loss. Despite inflation, the market price of existing houses in ravaged Detroit has hardly increased at all for a decade.

The quality of education provided by Detroit's public schools is widely regarded as low. In 1976, another 11,000 white pupils left the public schools, raising their ethnic mix from 75% to 79% black. The city's once extensive network of parochial schools has shriveled in recent years because of financial difficulties.

The bustle of shopping, office activity, and nightlife have moved to the suburbs where, at cocktail parties, one can hear matrons boast that they haven't been downtown since the 1967 riots. Crime has filled the wasteland left by the exodus of jobs and people, and in the early 1970s the city became known in urban folklore as "Murder Capital USA."[1] Downtown churches keep their doors locked except during services and taxi drivers warn visitors to the city against venturing out of their hotel alone after dark (conditions which, to be fair, are also found in many other troubled cities). During the summer of 1976, an epidemic of lawlessness struck Detroit. Roving gangs of teenagers assaulted and robbed stalled motorists on the freeways until state troopers took over patrol duties from city police. In August, one band of black youths went on a rampage, assaulting, robbing, and raping patrons at a downtown rock concert. Under intense pressure to halt the violence, Coleman Young, the city's first black mayor, recalled 450 laid off policemen, ordered a 10 p.m. curfew and vowed: "We do not intend to surrender the streets of the city." For most Detroiters, however, the streets had been surrendered long ago.

For all its problems, Detroit is still functioning and even fighting to recover. Belatedly, business leaders led by Henry Ford have made an extraordinary effort to revitalize downtown. Fifty-one companies chipped in $112 million of equity money and borrowed $245 million more to build Renaissance Center, a gleaming cluster of glass-walled office towers, shopping space, and a 70-story, 1,400 room hotel near the bank of the Detroit River. But the project has drained tenants from older downtown offices and its location, at the border of downtown, means that it is likely to provide little renaissance for the faltering heart of the city. More probably, Renaissance Center will become the nucleus of a new but smaller city center, patronized by whites, while the old center continues to turn slowly into a black shopping district.

Alone among large U.S. cities,[2] St. Louis had (by nearly 9%) a smaller population at the census bureau estimate of 1975 than it had in 1900—a fact that may help to explain why so many St. Louisans still speak with nostalgia about the greatness of their city "at the time of the fair," meaning the 1904 world's fair. The decline has been especially swift in recent decades. Between 1950 and 1975 the city's population shrank from 857,000 to 525,000. This 39% drop is the largest, in percentage terms, for any big American city during that quarter century. The abandonment of the city shows up even more strikingly in statistics compiled by the Center for Urban Programs at St. Louis University: since 1950 the number of dwelling units in the city has declined at a rate of nearly 2,000 a year, but the number of *occupied* dwelling units has declined at a rate of nearly 3,000 a year. By one recent count, a quarter of the city's population, including those who get food stamps, was receiving public assistance.

In great part, St. Louis's predicament is the consequence of its own foolish mistakes. The city has been hemmed into its 62 square miles of territory since 1876, when it shortsightedly cut itself free from the adjoining 497 square miles of suburban St. Louis County. At the time, St. Louis had little to fear from independence, for the city had been thriving ever since the days when its warehouses outfitted the pioneers and its location on the Mississippi River seemed to assure continuing prosperity. But since World War II, the county has passed the city in population and grabbed most of the area's fast-growing industry. In recent years, manufacturing, which has been the foundation of the St. Louis economy, has been declining. From 1969 to 1974 the number of manufacturing establishments in the city fell 18% (to 1,393) while the number in the metropolitan area as a whole dropped only 0.4% (to 3,106). The county seat of Clayton, which lies only 10 miles west of downtown St. Louis, has turned from a cow town into a financial, business, and apartment center that has plucked vitality out of downtown. Even the University Club of St. Louis is now domiciled in Clayton.

In its efforts to revivify old parts of the city, St. Louis has blundered again and again. Some land cleared with urban renewal subsidies has lain fallow for more than a decade. The high-rise Pruitt-Igoe public housing project, an architectural prizewinner when it opened in the mid-fifties, became loaded with welfare families and turned into a jungle of drugs and crime; despairing officials finally had wrecking crews flatten the most notorious section with dynamite. And downtown, despite its celebrated Gateway Arch, a sports stadium, and many a new office tower, generally empties out at 5 p.m. just like Detroit.

As rural migrants from the South poured into St. Louis, the black share of the city's population rose from 18% in 1950 to a locally estimated 43% in 1976. But St. Louis remained a sharply segregated city. The south side, with its blue-collar, ethnic neighborhoods of Italians, Germans, Dutch, and others, stayed almost entirely white. The north side–or what is left of it amid the boarded up storefronts, shard-strewn sidewalks, vacant lots and charred ruins–became almost entirely black occupied. Today it is a classic urban wasteland, inhabited by an underclass afflicted by unemployment, poverty, and crime. Having lost a substantial share of its middle-class white families to the suburbs in earlier years, St. Louis recently has been losing its black middle-class as well.

THE PROSPECTS FOR ADDITIONAL DOUGHNUTS

How many more doughnut-hole cities may the nation get? And how might that come about? To be sure, it takes more than population losses to turn a big central city into an economic and social slump in the middle of a bustling metropolitan area. But a large population decline is the first requisite. Since the end of World War II, most large central cities in the U.S. have been losing population steadily–except in the Sunbelt. Over the quarter century from 1950 to 1975 these losses add up to an astonishing total. Eleven cities–New York, Chicago, Los Angeles, Philadelphia, Detroit, San Francisco, Cleveland, Boston, St. Louis, Buffalo, and Minneapolis–lost more than 100,000 inhabitants each. Their combined loss was 3,122,000 people, a population larger than that of Chicago, itself the biggest loser (–521,571 people). Detroit's population dropped by 515,000, and New York City's by 410,000 (about as many as dwell in Cincinnati).

Eight cities lost a fifth or more of their 1950 population by 1975, among them the five whose economic fortunes appear most faded. At the top of the list was St. Louis, which lost 39% of its inhabitants, followed by Pittsburgh (–32%), Cleveland (–30%), Buffalo (–30%), Detroit (–28%), Minneapolis (–28%), Newark, N.J. (–23%), and Boston (–21%).[3] There are special reasons why Pittsburgh, Minneapolis, and Boston sustained such large population declines without a corresponding economic downspiral. All three have managed to retain lively and important downtown business districts. Pittsburgh has converted itself from a dark and filthy mill town into a corporate headquarters and research center. Minneapolis, thanks to its northerly location, was spared the influx of rural migrants who overwhelmed so many other cities. Boston, also an essentially "white" city, is an important financial center, the capital of Massachussetts, and the hub of the nation's largest cluster of colleges and universities.

Yet even for cities whose fortunes change with glacial speed under most circumstances, the key to the future is the *recent* past–that is, what has been happening in the seventies. By one recurrently propounded analysis, the economic and population decline of central cities has widened since 1970 into a metropolitan and regional phenomenon in the Northeast and North Central states. Some skepticism is in order. To be sure, 13 large metropolitan areas stretching from Providence, R.I., to St. Louis lost population through 1975. But the population declines in the central cities of 11 of those areas (and in all 13 combined) exceed the losses in metropolitan populations. It is statistically true that population growth has almost halted in the nine-state Northeast and that from 1970 to 1976 total population fell by 2.4% in Rhode Island and by 0.9% in New York State. But the entire decline in New York State population can be explained by the large drop in New York City. What is really going on, it appears, is that the accelerating decline of central cities is dragging down states and regions with them. But the real impact remains painfully concentrated in the same old places.

Since 1970, population decline in large central cities outside the Sunbelt has become all but universal. Among the top 50 cities in the U.S., only three outside the Sunbelt (Honolulu, San Jose, Calif., and Omaha) gained population between 1970 and 1975, while 25 were losers. The more surprising news is not that the 22 Sunbelt cities among the top 50 fared a lot better, but that they did nowhere nearly so well as the region's rising population and prosperity might suggest would be the case. There were 13 gainers and nine losers. Among the latter, Nashville and Oklahoma City lost a trivial 0.6% of their residents. Los Angeles, following the pattern for cities of more than 1,000,000 population (only exception: Houston), lost 3% of its inhabitants. But Dallas, where downtown dies at night and turns into a black shopping district on Saturdays, lost 3.7% of its population. Rival Ft. Worth lost 8.9%. New Orleans lost 5.7% of its residents, continuing a decline that began during the sixties. And Atlanta, for all its status as the commercial capital of the Southeast, lost nearly 12% of its population. Norfolk, Va., lost 6.9%.

These are ominous figures, for with the exceptions of New Orleans and Nashville (which also lost population during the 1950s), the declines appear to have reversed decades of population growth. For some Sunbelt cities, the rates of decline are comparatively high as well. In Atlanta the 1970-1975 rate exceeded that of both Detroit (by 0.1%) and Newark, N.J. (by 0.8%). Both New Orleans and Norfolk lost population at a faster rate than New York City (–5.2%).

THE RISE OF SUNBELT DOUGHNUTS

If these population trends continue to spread, it is possible that some Sunbelt cities, a decade or two hence, might find themselves with problems akin to those of Detroit or St. Louis. The key question is how the region handles its minority groups, both black and Mexican-American, which have shared little of the Sunbelt's broad economic advance in recent years. Numerous pockets of poverty remain, especially in black rural parts of the Deep South. In its Sunbelt expansion, industry has tended to build new factories where it can find surplus white labor and has avoided places with a high ratio of poor and unskilled blacks. Some companies follow that pattern because blacks are deemed more likely than whites to join unions; others are deterred because the educational level of blacks is lower than that of whites. Lacking economic opportunity in the countryside, the black poor in recent years have been migrating to big Southern cities. They cluster in expanding racial and economic ghettos, where their discontent is much more likely to find organized expression than in a rural, low density setting.

Some familiar symptoms or urban malaise have appeared. Atlanta, for example, has become a black-majority city surrounded by a fast-growing white suburban ring. The business-political partnership that built the city into a great regional center of commerce ended with the election of Atlanta's first black mayor, Maynard Jackson. As retail trade moves increasingly to the suburbs, the city has been struggling with a glut (perhaps temporary) of new hotel space, vacant offices, condominiums, and apartments.

New Orleans has been something of an economic backwater for about two decades, even though the city enjoys a flourishing tourist trade and has the nation's second busiest port (after New York). Since 1967, the number of industrial jobs in the area has declined every year but one. By 1977, only 11% of the labor force was employed in manufacturing, one of the lowest proportions among the nation's major urban centers. What economic development has occurred, notably an influx of highly automated petrochemical plants along the Mississippi River north of the city, has provided few jobs for the city's workers. In common with Mississippi and the rest of Louisiana, New Orleans suffers the consequences of long neglect of its schools: a large, unskilled underclass, mainly black. So many discouraged workers have given up even trying to find a job that the proportion of the New Orleans area population counted as part of its labor force is among the lowest in the nation: 55.5% in 1970 compared with 59.6% in all metropolitan areas, 64.7% in Atlanta, 65.7% in Dallas and 67.8% in Houston.

In short, some Sunbelt cities are beginning to repeat the misfortune of Northern cities by acquiring an increasing population of unskilled poor at a time when the industrial jobs those people need are moving out of town.

Several Sunbelt cities—among them Houston, Dallas, and Charlotte—have developed growth patterns that are economically and racially segmented. That is, well-to-do whites cluster in one sector, poor whites in another, poor blacks in a third, and, in some cases, poor Mexican-Americans in a fourth. In these circumstances, the rich and poor live and work in different worlds and seldom even see one another. In part this process reflects the normal workings of the real estate market. Yet because Sunbelt cities are being built at much lower densities than the older, industrial cities of the Northeast, Sunbelt cities will face increasing difficulties in providing community services to their sprawling low-income neighborhoods. It would not be surprising if, in a few decades, some Sunbelt cities develop large wastelands in which few will choose to live. There is, of course, time for most cities to forefend such an outcome, but whether they have the will and the skill is quite another matter.

THE CHANGING CONFIGURATION OF THE AMERICAN CITY

Given today's level of American affluence and technology, it is quite possible to have a metropolitan area *without* much of a central city. We no longer need very large cities to transact most business, and by their recent migration patterns Americans have shown an aversion to living in most of them. The rich and most of the middle class have the wherewithal to live where they choose and increasingly they have demonstrated a preference for more space, greenery, and tranquillity than is to be found in obsolescent, crime-plagued central cities. Large agglomerations such as New York, Chicago, and Philadelphia grew up in response to late 19th century technology and transportation. Factories, offices, and dwellings were jammed together for quick access to one another and to river and rail routes. But, today, old central cities have lost their historic locational advantage by having skills, markets, and manpower concentrated in one location. With the great shift to moving goods and people by truck and automobile, and the increasingly complex network of planes, phones, and computers that link distant places together, the population densities of our old cities have become anachronistic. Industry began decades ago to move out of cities in search of larger, cheaper sites for modern one-story factories. Today all kinds of urban activities, even such supporting services as advertising agencies, law offices, and consulting firms, have spread out at lower and more agreeable densities over ever-widening metropolitan regions–and beyond. For most Americans the resulting shrinkage of the big old cities is not calamity at all, for by moving away they have improved their way of life.

The proliferation of giant regional shopping centers, made possible by the new network of freeways around and through the old centers, has been particularly effective in siphoning vital elements of urbanity and economic muscle away from central business districts and scattering them around the periphery of large metropolitan areas. Many supercenters have turned into mini-cities of their own, attuned as downtown never was (and perhaps never will be) to the auto age. They begin with three or four major department stores and 100 or 150 smaller shops. As these outlets draw customers from a wide area, more elements of a city—offices, apartments, banks, stockbrokerage houses, motels, medical clinics, even churches—cluster around the expanse of asphalt that provides the essential free parking lot. One regional center in the Cleveland suburbs even includes a cemetery. Many centers have become the social, cultural, and recreational focal point of their communities. Developers go to considerable expense to turn their air-conditioned malls and huge enclosed courtyards into community gathering places. They sponsor such crowd-pulling events as beauty contests, senior proms, how-to-do-it demonstrations, and innumerable fashion, art, flower, furniture, boat, and auto shows. Suburbanites, a majority of the nation's metropolitan population since the 1960s, find the regional centers the convenient all-purpose place to go for anything from a Sunday stroll to a pop concert by a local symphony orchestra.

In the aggregate, the supercenters have wrought a remarkable transformation of urban geography. Their economic magnetism has split asunder the functions of many large cities. Instead of a single nucleus there are now several: the old downtown (which has frequently evolved into an office and financial district that dies at night), and a circle of smaller but often livelier satellite centers along the freeways five to 10 (sometimes 15) miles from the center of the city. The emergence of this new urban form is, if anything, more striking in the South and West, which developed more recently, than in the older Northeast. But the new shape, or elements of it, is to be found in every region of the nation. It seems fair to say, for instance, that the 66-mile Capital Beltway around Washington, D.C., has become the real main street of the metropolis. Over one recent 10 year span, some 800,000 new inhabitants settled along the beltway's borders; twelve regional shopping centers vied for their trade. Houston, Denver, and, to a slightly smaller degree, Atlanta all show similar patterns of development and the list could go on and on.

The trend horrifies some critics. "This free-for-all on the freeway guarantees the inner city's demise," complained Ade Louise Huxtable, the New York Times' Pulitzer Prizewinning architectural critic, not too long ago. That may be something of an overstatement, but it is true that the essence of the new urban form created by the super shopping centers—an intricate and compact orchestration of *mixed* land uses—is precisely what used to make downtowns lively and attractive. And that is the very quality that has been so often lost in the sterility of grandiose, single-purpose renewal projects in the inner cities.

The regional centers got built, one must remember, because private entrepreneurs figured (in most cases correctly) that it would be immensely profitable to build them. They were not conceived and constructed in order to transform the structure of our metropolitan areas. But that has been the result. And now that billions of dollars of investment have hardened the new urban form into steel, glass, and concrete we will have to live with the result for decades whether we like it or not.

For several years now, there have been efforts in scattered cities to revive retailing downtown, most often by the construction of bigger and better shopping centers. A few have proved moderately successful, but others have flopped or fizzled. The results have not inspired retailers to alter the suburban thrust of their expansion; almost all of it still goes into shopping centers on the edges of town. Downtown retailing seems likely to recapture its lost share of the metropolitan market only when and if cities become safer and more inviting places for families to live.

POLICIES FOR NORTHEASTERN DOUGHNUTS

Given the forces arrayed against their growth, big old cities must accept the inevitability of shrinkage in their economies and population and adjust to it. The alternative, as New York City discovered the hard way, is to become a ward of the state, deprived in fact though not in form of many essentials of self-government. That message has been slow to register. At a time when cities need skilled, tough, farsighted public leadership for a long sustained effort that might, over decades, revitalize them, all too many are led by weak, incompetent, or demagogic officials. Managerial ineptitude at city hall seems to be more common in the Northeast than in many other parts of the nation, perhaps because the city manager form of local government is *un*common. In any case, instead of retrenching and attacking their problems sensibly, a lot of cities have aggravated them by spending money they do not have, promising pensions that cannot be paid and showing an easy tolerance for swollen, underproductive municipal bureaucracies. New York City demonstrated all these traits during its spending binge of the late 1960s and early 1970s, which led to the thinly camouflaged state-federal receivership imposed in June 1976. Considering the structure of the city's government, the outcome may have been preordained. Felix G. Rohaytn, the Manhattan investment banker who has helped steer the city through the financial shoals as chairman of the state-created Municipal Assistance Corporation, calls New York City's government "the ultimate nonsystem" for managing a city. Rohaytn observes,

> The modern corporation is the last surviving form of autocracy. When somebody at the top makes a decision, not always the right decision, because information gets diluted as it travels up to the top, but it gets things happening down the line. But this city is like a Levantine negotiation. A decision is just the beginning of an argument about what's really going to happen. In the first place, the data isn't very good. People who study it don't always interpret it rationally. They may feel obliged to consider such questions as "Will this look like racial bias" or whathaveyou. So running the city is a process of cajolery and compromise. And by the time a decision gets down to the field where it has to be carried out, there's no way of knowning what's happening.

Still, New York has no monopoly on inept management. Cleveland, for example, sold its sanitation district to ease a money pinch and then spent the proceeds on operating expenses. Both Philadelphia and Boston flirted with the possibility of financial calamity early in 1977 because of overspending. Philadelphia faced such large unfunded pension obligations that Moody's marked its bond rating down to Baa. Boston got into financial hot water partly because some 2,000 workers were added to the city's payroll during Mayor Kevin White's first term and partly because the independent school board hired 1,483 new full- and part-time aides, probably a lot more than were necessary, following a federal court's controversial order to institute cross-city busing. Over the short run, such practices help politicians get reelected. But over the long run they raise the cost of government and force tax increases that prompt businessmen to look for localities where the burden of taxes and cost of doing business are lower.

Given first-rate local leadership, which is admittedly hard to come by, cities can cope with their inexorable shrinkage. Surprising as it may seem, there are at least two cases in point among the large cities in the troubled Northeast. Both have been burdened with such familiar woes as an increasing poor population, dwindling jobs, and an exodus of the middle class. Yet their resourceful leaders have found ways to cut expenses, curb crime, perk the place up with new action, and even to win battles with militant municipal unions. By a symbolic coincidence, in both cities a firm mayor personally led squads of volunteers picking up garbage during a successful effort to halt a strike by the sanitation union.

As mayor of Pittsburgh from 1970 until early 1977, Pete Flaherty showed an uncommon zeal for efficiency and economy. At a time when other cities were adding employees, he managed to reduce the municipal work force by 30%. At one point he also managed to reduce the city's tax take by 20% and even with persistent inflation the tax load in 1976 was 3% below the 1970 level. In sparing Pittsburgh a financial quagmire, Flaherty managed at one time or another to make an enemy of almost every powerful group in town: organized labor, most of the city council, police and firemen, the business community, the black community, and both daily newspapers. Critics called him callous, cruel, ungrateful, regressive, arrogant, and obstructionist. But the voters loved him. He was reelected overwhelmingly in 1973 without the support of the unions, the Republicans, or even his own Democratic party. Facing considerable opposition for a third term in 1977, he joined the Carter administration as Deputy U.S. Attorney General.

Cutting back was an urgent matter for Pittsburgh. Between 1950 and 1975, the city's population declined 32% to 458,651, the second sharpest drop (as noted earlier) among the nation's leading cities; for several recent years, five dwelling units (mostly abandoned) have been demolished for every new one built. Yet, when he took office, Flaherty found the same conditions that beset cities all over the Northeast: inefficiency, swollen payrolls, and duplication of effort. In his first year as mayor, he replaced nearly all of Pittsburgh's department heads, froze hiring, placed restrictions on the use of overtime, and eliminated 900 jobs. That converted a threatened $2,000,000 deficit into a $3,700,000 surplus.

A fight with municipal unions broke out the following year after Flaherty wiped out five drivers' jobs in his proposed budget. The drivers, members of Teamsters' Local 249 (whose president was also a member of the city council), had been chauffering pickup trucks for plumbers who installed residential water meters. The plumbers, however, refused to drive for themselves and after the Teamsters

went on strike the walkout grew to involve more than 2,000 municipal employees including the plumbers, street crews, mechanics, and garbage collectors. Flaherty went on television and accused the unions of "trying to take over the city." Then, with TV cameras still following him, he led a volunteer trash pickup squad. Ten days later the unions called off their strike.

After that confrontation, Flaherty relied on attrition to pare the city's work force. "You have to cut back long term, or there are too many union problems," he observed. Far from crippling municipal services, the manpower cutbacks seemed to improve efficiency. "The city government works much better than it did with more people," Flaherty insisted. To be sure, Pittsburghers griped about laggard garbage collection and the condition of the streets. The mayor argued that many of the bumpiest thoroughfares were state highways and along such routes the city cannily erected signs reading: "State-maintained road." In any case, potholes and pinch-penny government seem a reasonable price for Pittsburgh to have paid to avert a financial debacle and enjoy an A-1 Moody's rating on its bonds. Reported crime dropped slightly, partly because of a street lighting program that contributed to a new after-dark bustle downtown.

In many ways, Pittsburgh's downtown Golden Triangle is a model of what an old city can do to retain a vital role in a large metropolis. Despite a great deal of new construction, including six new office skycrapers in a decade, the close-grained diversity of land use so essential to the fabric of any downtown has been preserved. The place is alive, complete, and marvelously compact (it is possible to walk from one end to the other in about 10 minutes). The streets are safe. Retail sales have been rising. Swinging singles jam the bars around Market Square. Though a lot of industrial activity has moved to the suburbs, increasing white-collar employment downtown has enabled Pittsburgh to do a lot better than many other old cities in maintaining total employment, which declined by only 5% from 1971 to 1976.

From his ballroom-sized office in one of the shabbiest city halls around, Pete Flaherty likes to admire the shimmering new corporate skyscrapers. Their presence, though, is the result of the celebrated "Pittsburgh Renaissance," the downtown rejuvenation fashioned by the durable (1945-1969) coalition between industrialist Richard King Mellon and former mayors David Lawrence and Joseph Barr. Mellon died soon after Flaherty became mayor and the alliance withered. It needs to be put back together. Pittsburgh can scarcely hope to expand, but with revived government-business teamwork, the city might at least build on its attractiveness as a regional commercial center.

Like the bumblebee, which theoretically should be unable to fly, Baltimore theoretically ought to be a city inextricably caught in a vortex of decline. It is an old, conservative, blue-collar industrial metropolis and it has been in trouble for decades. Back in 1927, H.L. Mencken wrote to his friend James Thurber: "I hope you drop off in Baltimore some day and let me show you the ruins of a once great medieval city." Baltimore remains the nation's seventh largest city, but the population has been dwindling since 1950; it fell another 6% from 1970 to 1975 (to 851,698) and the number of persons at work in the area declined at twice that rate. More than half the city's residents are black and the ratio of poor is high. Yet the city has made such an extraordinary comeback that it just might, as some Baltimore enthusiasts insist, turn into the queen of the Eastern seaboard.

Baltimore's revival, though far from total (downtown retail sales have continued to weaken), is a triumph of common sense over pitfalls in the urban renewal process that has gone awry in so many other cities. The most important ingredient is 20 years of teamwork between local officials and private business leaders. From the beginning, they followed a strategy geared to the fundamental but often overlooked fact that the protean force that shapes cities is self-interest—of individuals and companies. Accordingly, they set out to convert the heart of the city into a culturally rich, architecturally enticing magnet that would attract both affluent and middle-class families as a place to work, shop, and live. While retaining legal control, the city wisely placed the $1 billion job in private managerial hands—the nonprofit Charles Center-Inner Harbor Management Corp.—thus avoiding the creation of a municipal bureaucracy with all its red tape and torpor. Charles Center, an attractive 33-acre complex of office buildings, apartments, a hotel, shops, and tree-lined plazas, has replaced the dingiest part of downtown. And the adjacent 240-acre Inner Harbor has been transformed from a tangle of rotting piers and ancient warehouses into an enticing blend of parks and playgrounds, tourist attractions, restaurants, and office buildings.

Like the developers of successful shopping centers, Baltimore has also worked hard at making Charles Center and Inner Harbor the cultural and recreational focal points of the community. A galaxy of special events arranged mainly by a city-sponsored Downtown Coordinating Office draws crowds, including suburbanites, to the center of town at night and on weekends. Besides such conventional attractions as symphonies, ballet, a civic opera, and live theater, from May through October there are outdoor boat and craft shows, regattas, festivals, canoe races, and art exhibits along with almost endless band, choral, folk music, and jazz concerts (but no rock concerts, which tend to attract unruly customers). The biggest event, a three-day city fair at the end of September 1976, was attended by some 1,800,000 persons.

Baltimore's metamorphosis has been deftly managed since 1972 by a remarkable mayor, William Donald Schaefer, who has been able to impose stringent economies in government and tame the municipal unions without making a lot of political enemies. A hiring freeze that he imposed during this first year in office helped the city to gradually reduce its payroll by 2,200 employees, to 27,000. Schaefer continued the city's tight-fisted policy, dating from 1960, of limiting its borrowing to $35 million a year—all of it for capital projects. While inflation led most cities to increase their borrowing, Baltimore from 1972 to 1976 retired more bonds than it issued.

Schaefer's battle with municipal unions came in 1974 when police, city prison guards, and sanitation workers struck simultaneously. While the mayor and most of his administrative aides manned garbage trucks, winning a good deal of public sympathy, the police commissioner fired 75 patrolmen for spurning a back to work order. The unions and some of their officers were fined a total of $157,000 for defying court injunctions against the strike, and the police lost their right to bargain collectively. After that setback, the municipal unions in 1977 signed a two-year contract giving them a modest 4% annual pay increase. Many Baltimoreans contend that garbage collection improved after the strike. Schaefer had a smiling face painted on each trash truck and beneath it the admonition: "Have a happy day." To humor people into tossing litter into receptacles instead of gutters, the city's trashbaskets now carry a sign reading: "Throw a trashball." It sounds corny, but it seems to work. Once dirty Baltimore today has become a tidy looking place.

Numerous problems persist, of course. The crime rate, though declining, is still quite high. Large parts of the inner city remain black slums. The exodus of the middle class to the suburbs has diminished, but not ended—a condition that some Baltimoreans blame on the forced integration of public schools. Some 70% of the city's schoolchildren are black and federal administrators forced the school board to establish a 70% black, 30% white ethnic mix in every school on pain of losing $25 million a year in federal aid. In the fall of 1976, Baltimore school enrollment fell 4% from the level of the year before, a larger decline than can be explained by falling birthrates.

REFILLING THE METROPOLITAN DOUGHNUT

At a time when most other cities in similar circumstances were slipping backwards, it was no small achievement for Pittsburgh to have, in effect, managed to stand still during Mayor Flaherty's regime. For Baltimore to have moved forward, if only a bit, was a civic miracle. Both cities, of course, worked with the tools and politics at hand, a lot of which are completely inadequate if not actually counterproductive.[4] If central cities are to make any general comeback, if those holes in the metropolitan doughnut—present and prospective—are to be refilled, a good many radical changes in policy will have to be adopted, including some that are not yet being discussed even among experts.

The failures of our failing cities (and states and regions as well) are commonly viewed as economic, but their roots are quintessentially political and governmental. Enduring remedies will require political and governmental action that cuts across a vast array of entrenched beliefs and vested interests in waste, patronage, and job security. If we do it at all—a big if—restoring our big but troubled cities to tranquillity and economic health will take not years but several decades.

The federal government can help the most by resisting actions that damage the financial health of cities and retard their natural regeneration. In particular, Washington should give top priority to reducing inflation, which makes cities' costs (about two-thirds wages and fringes) leap ahead of their revenues. But a big infusion of fresh federal aid or, as the U.S. Conference of Mayors likes to call it, a "Marshall Plan for Cities," would be a mistake. In the first place, it would be inflationary and for the reason alone self-defeating.

Moreover, states and localities already receive some $70 billion a year in federal aid, a large but unquantifiable amount of which is wasted. More "free" money from Washington would give too many local officials an excuse to postpone once again making difficult decisions such as firing excess employees or resisting exorbitant union demands at the bargaining table.

If central cities are to regain a measure of their former vitality, it will be up to the cities themselves (with help from state legislatures) to do most of the job. For the great tragedy of our troubled cities is not their shrinkage but the fact that they have promoted their own decline and decay by conducting, or at least condoning, a form of economic warfare against their own interests. Laws, programs, and policies promulgated by government have created an array of perverse incentives that give both individuals and companies a self-interest in doing what is bad for the community as a whole.

Taken individually, these disincentives may appear as trivial as parking beneath a "no parking" sign. In the aggregate, they are not trivial—not even illegal parking, which is a major source of the traffic congestion that helps to depress the retail sales so vital to a city's economic health. Illegal parking persists, of course, because motorists correctly figure they come out ahead paying an occasional fine than a regular fee to garage their cars. Cities are adept at simplistic pricing practices that have unintended evil side effects. Municipal garages all too often have a schedule of charges rigged in favor of all-day parkers. That is popular, of course, with people who work nearby but it hurts retailers whose customers must pay dearly to park two or three hours, and who may be further deterred from a major purchase by a local sales tax not imposed in the suburbs. Most cities set prices on sewer and water services in a way that fails to reflect the varying cost of installation. Thus customers at the far end of costly long lines pay the same rates as customers who use older shorter lines. By such a combination of overpricing and underpricing, cities put an extra burden on old inner neighborhoods which, presumably, they are trying to revitalize or save.

In contrast with these small—and little noticed—disincentives, rent control is a particularly blatant and widely recognized scourge. It discourages the construction of new apartments and inhibits the maintenance of old ones. In New York City, where controls have been in effect continuously since 1943, they have led to widespread tax delinquency, premature abandonment of thousands of structurally sound buildings, and the decay of entire neighborhoods. Though rent control has come to be widely condemned in recent years as unsound public policy, the New York legislature remains too fearful of the wrath of well organized tenant groups to repeal it. And bizarre though it may seem, other localities have been copying the New York formula for a disaster in housing. During the 1970s. rent control has been adopted in some 100 New Jersey communities, Boston and three suburbs, Miami Beach, Washington, D.C., and Montgomery County, Md.

By far the most important way cities harness the profit motive backwards to their own detriment is through ill-conceived taxes,[5] which give both businesses and the affluent an incentive to leave. Manufacturing appears to be especially sensitive. In a report issued in December 1976, New York's Temporary Commission on City Finances identified manufacturing as "the major weakness in the city's economy" and blamed that weakness in great part on a punishing tax load. Between 1950 and 1975, the report noted, the city lost 500,000, or half, of its factory jobs. Unless taxes are greatly reduced, the commission warned, the city may lose another 150,000 manufacturing jobs by 1981, with a consequent drop of $500 million in revenues.

It seems fair to say that our cruelest use of perverse incentives involves our treatment of the poor. In the name of "helping" them we load them with subtle chains that tend to perpetuate their dependency. We deny them options, including the option to be productive, when we raise minimum wages and thus wipe out an important number of low-skill jobs. If employers pay workers with low productivity a high wage, they are apt to lose money, so the easy alternative is not to hire such workers. Work, as Freud recognized, is a powerful integrator of the human psyche. But America says to its unskilled jobless: "Society has no use for you." So some unemployed come to have no use for themselves—or society.

Compounding that perversity, welfare benefits have been set so high in some states as to create an incentive for not seeking work. Persons on welfare sometimes turn down proffered jobs in justifiable fear that they may soon lose them, in which case they will face a wait of several weeks to get back on welfare. Subsidized housing, on top of its other defects, has an insidious perversity that is not widely recognized. When a family of working poor moves into such accommodations, it loses the mobility that the breadwinner might need to take a better job in another locality. The reason is simple. A family that gives up a subsidized housing unit in one place has no automatic right to claim another one in another community. If the family moves, it must wait its turn for such quarters in its new location. The queue is often a long one.

It is, of course, harmful to the fiscal health of cities and, therefore, in the long run to their economies, to harbor disproportionate numbers of the poor. Cities depend on the property tax for 82% of the $61 billion a year they raise from their own taxpayers. But the value of real estate, and therefore its tax yield, tends to rise in prospering areas and fall in declining ones. The more aid cities provide for the poor, the more of them they invite to congregate within their borders. As their numbers increase, affecting property values, the cities' principal source of revenue is undermined. And when cities or states use their powers to redistribute wealth, as for example by combining high income taxes with generous welfare payments, they simultaneously create incentives for dependent elements of society and disincentives for the productive elements.[6] By and large, Americans simply do not understand what makes their cities function. One result is the triumph of misguided altruism over economic common sense.

The burden of poverty in U.S. cities, to be sure, has been considerably aggravated by the great migrations in the decades since World War II. Millions of rural poor, mostly black at first but more recently heavily Hispanic, poured into Northern and Midwestern cities in the hope of self-betterment that for many proved to be illusory. To be sure, U.S. cities have long harbored large numbers of impoverished immigrants in crime-ridden slums, as for example when they were absorbing newcomers from Ireland, Italy, or Eastern Europe. But in recent years the disappearance of factory jobs they once might have held plus the combination of increasingly generous welfare and the disastrous concentration of subsidized housing in deteriorating neighborhoods has tended to trap them in inner cities.

At the same time, our efforts to end racial segregation in public schools have had, in many cities, an unintended reverse effect. When the U.S. Supreme Court handed down its landmark school desegregation decision in 1954, President Eisenhower ordered the immediate integration of the Washington, D.C., public schools and was widely praised for his action by liberal voices across the country. Thereafter, white families pulled their children out of the District of Columbia schools at an accelerated rate. Now that the public schools in Washington are 95% black, the process of running away has spread to adjacent Prince Georges County, Md., where many black middle-class families moved to avoid the crime and violence of all too many neighborhoods in the capital. In the four years ended in 1976, white enrollment at Prince Georges's schools fell by 25%. Cross-city busing, the federal courts' more recent and highly controversial weapon against school segregation, is prompting more middle-class families to leave cities at a time when cities need to hang on to every middle-class family, black or white, that they can.

This pattern of white flight, and the segmented development patterns to be found in some Sunbelt cities, are two visible facets of an ominous trend; we are building an increasingly apartheid civilization despite our morally just equal rights laws. The frozen political boundaries of most inner cities, which prevent them from annexing more territory and tax base as a way out of their fiscal crisis, reflect suburban determination to avoid cities' social and racial problems (and tax burdens). Fewer and fewer businessmen seem willing to maintain retail outlets inside big-city ghettos and those that do generally feel forced to raise their prices to compensate for increased shoplifting, higher insurance premiums (if they can get insurance at all), and other hazards. A senior executive of one of the nation's top retail chains explained the dilemma to me with a candor born of several years acquaintance. "You can't get an able store manager to work in a low-income area, least of all a black manager," he said. "He'll get ripped off worst of all. They tell him: 'You're working for whitey.' You have to nail everything down and shift to a lower grade of merchandise. So there is almost no store expansion in the inner cities. The black middle class goes out to suburban shopping centers, and I don't blame them."

Lamentably, the national habit of thinking about blacks and other minorities as ethnic symbols rather than as individuals makes these problems more intractable. As in years past, most of today's newcomers to inner cities in time will move up the economic ladder. But some will not. The real problem involves a small lower class, perhaps only 15%, whose tendencies toward violence, dislike of steady work, and inability to maintain a steady family life keep them permanently mired in poverty. But as Harvard's Edward C. Banfield pointed out in his seminal book, *The Unheavenly City,* nobody has learned how to change cultural habits any faster than time or circumstances do so naturally. A more draconian approach to this problem, which understandably is not on the national agenda, was suggested three decades ago by the late Harvard anthropologist, E.A. Hooten. In a talk at Stanford University, he declared: "If we must feed and foster the shiftless and incompetent, we must at least prevent them from reproducing their kind, or else democracy is lost." As a society, we have not yet begun to consider, let alone decide, what solution if any is appropriate for irresponsible fecundity. But the problem is a real one, and our troubled cities bear the brunt of it.

CONCLUSIONS

With prescient insight, the British historian Macauley wrote about a century ago that, unlike imperial Rome, America seemed safe enough from destruction by invading barbarians from beyond its borders, but had much to fear from barbarians within them. As the thousands of Americans who in recent years have been mugged, raped, shot, stabbed, robbed, beaten, or burglarized might bear witness, an alarming number of barbarians are on the loose in our cities. New York's south Bronx, a neighborhood with a population (last time it was counted) of 450,000 (which made it more populous than Atlanta), has been systematically laid waste by crime (notable arson) that is mainly the handiwork of its own inhabitants. Many New Yorkers had figured that the south Bronx was a special case, but the orgy of looting and arson during and after a citywide power blackout on the hot muggy night of July 13, 1977, provides ground for taking Macauley's prophecy at something close to face value. In 16 black and Hispanic neighborhoods of the Bronx, Brooklyn, Manhattan, and Queens, looters plundered some 1,300 food, liquor, clothing, appliance, jewelry, and other stores. Some were even stripped bare of their fixtures. The maurauders moved almost as though on signal at the start of the power failure and, sometimes working in gangs of 20 or 30, continued brazenly into the daylight hours of the following day. True to the city's permissive form, the police did not try to stop the plunder by force, though they did arrest 4,471 persons (3,841 of them for looting). "It's the night of the animals," said one weary police sergeant. "You grab four or five and a hundred take their place."

Crime has proliferated in our cities because it is the most rewarding way of life available to those involved, notably jobless teenagers with low skills. Today more than half of all serious crimes (murder, rape, aggravated assault, robbery, burglary, larceny, and motor vehicle theft) are committed by youths aged 10 to 17. Since 1960 juvenile crime has risen at twice the rate of adult crime. The miscreants, of course, come from every kind of environment and ethnic group, but about half of violent juvenile crime is committed by black youths, and a large but unquantifiable amount by Hispanics.

Cities, and the civilized values they represent, will not permanently endure if they continue to cower before the depredations of habitual criminals—adult or juvenile. The explosion of leniency in our courts has, if anything, made matters worse. To increase public safety, the foremost task of government, it is time that repeaters and violent criminals including juveniles be imprisoned for lengthy terms. Our overcrowded courts and prisons would surely be expanded at a small fraction of the cost borne by urban businessmen and residents to fend off crime or flee from it. If ailing central cities are to be saved as centers of commerce and culture, there is no more urgent task than making their streets safe. To be fair about it, we will have to find ways to increase the opportunities for work as we increase the punishment rate for crime. Public service jobs, the solution favored in Washington, have so far proved to be an ineffective way to instill work habits in high school dropouts. But until somebody devises a sounder approach that is also politically feasible, a palliative may be better than nothing.

The second urgent task facing cities in search of revitalization is to improve the quality of education in their public schools. This, too, will require radical surgery on the system itself. At present, we have coercion for pupils, who are required to attend even after they have shown they can no longer profit from formal schooling, and a no-incentive system for educators. Teachers are usually paid according to seniority plus the number of degrees they happen to have, not according to how ably they perform their vital task of teaching. Middle-level administrators have no more incentive to be efficient than other civil servants (whose remuneration often depends on how many underlings they supervise), and top administrators find they must defend the system or lose control of it. Is it any wonder that the more money we pour into schools, the more illiterates seem to pour out of them? A free public education is a costly privilege, not to be denied the majority by supine tolerance for incorrigible classroom disrupters, who ought to be expelled. Toward that end, the age for mandatory school attendance should be reduced, perhaps to 15 and maybe lower. For the long run, more fundamental change is needed, for example harnessing education to the profit motive. Economist Milton Friedman's proposal for a voucher system, subsidizing the customer instead of the bureaucracy, is logical and sensible. It would force every school to provide an education that satisfies public demand or, like a mismanaged supermarket, lose its clientele. With their jobs at stake, teachers and administrators at last would have a real incentive to perform.

A similar restructuring of incentives—no less difficult to achieve than the kind of school reform Friedman envisages—would go a long way to remedy the managerial shortcomings in local governments. Civil service laws and regulations must be overhauled to make it much easier to fire slothful or incompetent employees, and to give middle level management a reason to put efficiency above self-protection. Ultimately, something must be done to curb the awesome power of municipal unions or cities, like some school systems, will be run primarily for the comfort, convenience, and security of their bureaucrats rather than the benefit of the public.

Finally, if our sick cities are to regain their economic health, they will have to learn to compete. A surefire way to drive a city into economic decline is to let costs and taxes get way out of line with costs and taxes in other places, as has happened in New York. Business and industry will always gravitate to these parts of the nation where it is cheaper to operate. Right now that means the Sunbelt, partly because wages run a bit lower, partly because goods and services are a bit less expensive, but mainly because of a lower burden of taxation. In the fall of 1975, according to the Bureau of Labor Statistics, it cost $27,071 in the New York metropolitan area and $27,000 in the Boston area for a family of four to maintain the same "higher" standard of living available for $20,090 around Houston and $20,362 around Atlanta. By my estimates the New York family would have paid total taxes of $11,038 (including taxes hidden in the price of goods and services); the Houston family would have paid only $5,981 in taxes. The largest part of the difference comes from state and local taxes.

It is logical to expect the great disparities in costs between the Sunbelt's "cheap" cities and the Northeast's costly ones to narrow over time. Prices and wages have recently been rising at a faster rate in the Sunbelt than in the nation as a whole. However, so far the dollars-and-cents gap between Houston and New York City, for instance, is so large that Houston's faster *rate* of inflation does not seem to be reducing it.

The future for our ailing old cities looks precarious now, but 20 or 30 years hence it may look better. Much will depend on the future cost and availability of energy (especially petroleum), for the economy of the South and West (and their cities) is both more energy-intensive and more auto-dependent than that of the Northeast (and its cities). It is not beyond possibility that severe new federal restrictions on petroleum use, or merely another huge rise in the price of oil, would strip the Sunbelt of the economic advantages it has been enjoying.

In any case, it is a fallacy to suppose that cities, like living organisms, inexorably follow life-like cycles of growth, aging, and decline. The fortunes of cities, like that of the stock market, fluctuate. As economist Miles L. Colean wrote a quarter century ago in his book, *Renewing Our Cities,* "like all creators man creates in his own image and, if he does not like the reflection he finds in his city, he has his own lack of interest, competence or foresight to blame. If his city is confused or disorderly, that is because he himself is confused or unmindful of disorder, or lacking in the means to achieve order. . . . If the city sinks into decay, it is because he has lost the wish or vigor to maintain it."

Our cities will be what we make them and we have been unmaking them with foolish public policies, notably those that hitch the profit motive to the wrong objectives. Many of the most serious difficulties—crime, bad schools, the civil service bureaucracy, even managerial deficiencies at city hall—should be lessened if those perverse incentives are turned around. Since most of the necessary changes would impinge on the special rights and privileges of well organized and powerful factions, the chances of their adoption look very poor right now. But nothing lies beyond change and what is politically unthinkable today may be conventional wisdom tomorrow. For the moment, we will no doubt rely on expedient half measures, not meaningful reform. In time, with wider understanding of what makes cities prosper or decay, we may get wiser laws and abler political leadership. Cities and the people in them are resilient, and they have demonstrated throughout history an extraordinary ability to adjust to new circumstances. They will continue to adjust.

NOTES

1. During the first half of 1977, crime in Detroit declined in every major category. The murder rate fell 28% compared with the first half of 1976.

2. Providence, R.I., the nation's 86th largest city in 1975, also experienced a decline in population over the first three-quarters of the 20th century. The population dropped from 175,597 in 1900 to 167,724 in 1975.

3. Several cities whose 1975 populations place them below the 50th rank also had large decreases in population from 1950 to 1975. Providence, R.I., lost one-third of its inhabitants. New Haven, Conn. (no. 117), lost 23% of its populace; Hartford, Conn. (no. 108), lost 22%.

4. The shortcomings of federal programs aimed at urban problems have been evident to perceptive analysts for many years. For four decades Congress has been piling one subsidy on top of another, beginning with subsidized public housing and running on through the so-called "war on poverty" (which was declared but scarcely fought) to revenue sharing in a piecemeal attack about as effective as pushing wet spaghetti. Apparently learning little or nothing from prior mistakes, our legislators repeat them in fresh disguises. One measure of the stagnation of thought about urban problems is that even decade-old protests at the futility of federal efforts sound as though they were fresh. For example, in a 1968 lecture at Boston University, Edward J. Logue, one of the nation's leading urbanologists, said: "Our present array of urban policies and practices are not going to solve the urban crisis and simply beefing them up will not work either. When it gets down to the day-to-day realities of urban policy implementation, it is C. Northcote Parkinson who sets the rules, not the Congress, not the White House, nor even the top brass at the Department of Housing and Urban Development." As Logue saw it: "Under the present rules of the urban game only exceptional mayors have a prayer of making a significant impact. We must design systems for dealing with the urban crisis which do not require so much talent because it is in too short supply."

5. The property tax may well be the most ill-conceived tax of all, because of the harm it inflicts on cities by fostering the formation and perpetuation of slums. The tax is a confusing and little understood amalgam of two separate levies, one on the building and the other on the value of its location. Most cities collect two or three times as much tax from buildings as from the site value of the land. The low levy on land makes it easier for speculators to keep idle or underutilized parcels off the market until (they hope) population and economic growth enable them to sell such property at a big profit. The high levy on buildings (or improvements to them) penalizes the construction of new buildings and the maintenance of old ones—exactly what cities do not need. Implicitly recognizing the shortcomings of the realty tax structure, numerous cities offer special tax concessions to induce entrepreneurs to erect major new buildings that are considered desirable by whatever faction is in power locally. This is, of course, unfair to all other taxpayers. The property tax should be turned around. The levy on buildings should be greatly reduced or, better, abolished entirely; a corresponding and offsetting increase in taxes should be imposed on land. Such a reversal would increase the incentive for private investments in good buildings, as distinguished from the schlock invited by our present tax arrangements. And it should reduce the profitability of land speculation, which is, in many ways, an anti-social activity.

6. New York State and City provide a prime example of this form of cooperative urbicide by taxation and lavish welfare spending. In its final report issued in May 1976, the Special Task Force on Taxation, organized at the request of Governor Hugh L. Carey, noted that New York for more than a decade has imposed the highest state and local taxes per capita in the U.S. In 1974, the burden was $952.29 per person, 25% higher than the burden in the next two states, Hawaii and California, and 55% above the national average. Said the report: "The tax burden currently imposed on personal and business incomes, particularly

in New York City where state taxes are duplicated and extended by local levies, has risen to a level totally out of balance in comparison with the burdens imposed by our neighboring states, in the whole of the northeast region and, indeed almost everywhere in the nation." There is, of course, a reason for such high taxation. New York City not only spends more money per capita ($1,382 in 1974) than any other major city in the country except Washington, D.C., but allocates some 40% of that spending for social services, about double the national average. Recently the city has been funneling some $4.8 billion a year into wefare, health, and hospitals alone. Having cited most of these figures, the Task Force report warned: "A sizable and growing number of our more prosperous citizens perceive life in New York City as a deteriorating condition. Against this declining return on the country's highest tax cost they must weigh the substantially lower taxes and often substantially better school and community services offered by neighboring and more remote states and localities. It is hardly surprising that New York is no longer a Mecca for rising young executives and that more and more of those who are already here shift their business or profession to other states if they are able to do so. This flight of our most productive citizens will continue unless and until the unfavorable balance of costs and benefits of living and working in New York is reversed."

11

Toward a Vision of the Urban Future

MURRAY BOOKCHIN

□ "WITHOUT TESTAMENT," observed Hannah Arendt (1954:6), "without tradition—which selects and names, which hands down and preserves, which indicates where the treasures are and what their worth is—there seems to be no willed continuity in time and hence, humanly speaking, neither past nor future, only sempiternal change of the world and the biological cycle of creatures in it."

If the city can be added to the lost treasures which Arendt laments in her deeply sensitive essays, this loss is due in no small measure to the modern stance of "contemporaneity," a stance which virtually denies an urban past in its deadening claim to sempiternal change, to an eternality of problems that have neither the retrospect of uniqueness nor the prospect of visonary solutions. To the degree that the very word "city" is still applied to the formless urban agglomerations that blot the human landscape, we live with the shallow myth that the problems of the civic present are equatable with those of the civic past—and hence, in a sinister sense, with the civic future. Accordingly, we know neither past nor future but only a present that lacks even the self-consciousness of its social preconditions, limitations, and historic fragility.

Our very language betrays the limitations within which we operate—more precisely, the preconceptions with which we define the functions of the modern city and our "solutions" to its problems. However operational it may be, the most unspoken preconception that guides our view of the modern city is an entirely entrepreneurial one. Indeed, all shabby moral platitudes aside, we simply view the city as a business enterprise. Our underlying urban problems are commonly described in fiscal terms and often attributed to "poor management," "financial irresponsibility," and "imbalanced budgets." Judging from this terminology, it would seem that a "good city" is a fiscally secure city, and the job of civic institutions is to manage the city as a "sound business." Presumably, the "best city" is not only one that balances its budget and is self-financing but even earns a sizable profit.

To anyone who has even a glancing acquaintance with urban history, this is a breathtaking notion of the city, indeed, a notion that could arise only in the most unadorned and mediocre of bourgeois epochs. Yet lacking a sense of both past and future, we would do well to note that the city has variously been seen as a ceremonial center (the temple city), an administrative center (the palace city), a civic fraternity (the polis), and guild city (the medieval commune). Heavenly or secular, it has always been uniquely a social space, the terrain in which the suspect "stranger" became transformed into the citizen—this, as distinguished from the biological parochialism of the clan and tribe with its roots in blood ties, the sexual division of labor, and age groups. As the Greeks so well knew, the "good city" represented the triumph of society over biology, of reason over impulse, of humanity over folkdom. That capitalism with its principle of unlimited growth and its economic emphasis on "sempiternal change" began to expand the medieval marketplace beyond any comprehensible human scale is a problem that has been more than adequately explored; but where this tendency would take us was still conjectural.[1] The last century saw the city defer to—and even model itself—on the factory (Bookchin, 1973:51-52, 89-92). The opening years of the present century witnessed the conceptual reduction of the city to a "machine," a notion which was accepted by such presumably disparate architects and planners as Le Corbusier (1971:1) and Frank Lloyd Wright (1964:94). McLuhan brought us into the multinational corporate world with his catchy phrase "the global village" (1968), and Doxiades presumably afforded us the "multidisciplinary" tools for making the multinational city seem like an international one (1965).

If the schemes of Le Corbusier, Wright, McLuhan, and Doxiades seem remote at present, if they have been preempted by the bookkeeping of Abe Beame, the shift is not without its irony. Beame plodded his way to the center of New York politics as a comptroller, not as a social reformer or city planner. His concept of community is probably exhausted by the New York Democratic Party's headquarters and backrooms. He lacks even the Dickensian eccentricities of a Scrooge. Only his gray hair, aging face, and diminutive stature rescue him from appearing as a corporate technocrat. He is, oddly enough, a man of the LaGuardia generation who, like the Abbe Sieyes of the French Revolution, could claim a supreme credential for having lived in a colorful, dramatic, and dynamic era: he managed to survive. By virtue of his very appearance and professional background, Beame personifies the transformation of New York's urban problems from those of social reform into those of fiscal manipulation.

Lest this transformation be taken too much for granted, it has implications that go far beyond any mere headlines. The change means that our modern capitalist society has not only subverted the city's historic role as a medium for socializing parochial fold into worldly humans; it has completely degraded the city into a mere business venture to be gauged by monetary rather than social or cultural criteria. It has, in effect, added a vulgar dimension to Arendt's worst fears of "sempiternal change" by removing the city from the history books and placing it in account books. The city has become a problem not in social theory, community, or psychology but in bookkeeping. It has ceased to be a human creation and has become a commodity. Its achievement is to be judged not by architectural beauty, cultural inspiration, and human association but by economic productivity, taxable resources, and fiscal success. The most startling aspect of this development—long in the making when the city was subordinated to the factory and to commerce—is that urban theory must cease to pretend that its revered social and cultural criteria apply to the modern city. Architecture, sociology, anthropology, planning, and cultural history tell us nothing about the city as it exists today. Urban ideology is business ideology. Its tools are not Doxiades's ekistics but double-entry bookkeeping.

The extent to which we have removed the city from the history books and placed it in the account books is evidenced not only by the declining cities of the Northeast but by the burgeoning cities of the Sunbelt. Success here is a quixotic form of failure–for the historic urban trend of our days has not been toward cities but rather a curious form of urbanization without cities (Bookchin, forthcoming). The devolution of the Sunbelt cities almost entirely into industrial and commercial "mousetraps" (to quote a *Fortune* journalist) has yielded a devastating form of "success." Business has become a cult; growth, a deity; money, a talisman. The mythic has reappeared in its most mundane quantified form to create one of the most dehumanizing ideologies in urban history. In the plastic, unadorned subdivisions, high-rises, and slab office buildings of Los Angeles, Phoenix, Dallas, and Houston eastward to Tampa and Miami, life and culture have been scarificed to the most robotized forms of mass production, mass merchandizing, and mass culture. The faceless structures that sprawl across the "southern rim" lack the seasoning of history, of authentic cultural intercourse, of urban development and centering. The cities themselves have moved, for the most part, by huge leaps, not by evolution, and the propelling force of the leaps has been some sort of "resource," be it copper or petroleum, aerospace or electronics, range empires or agribusiness. The "gold fever" has never left the Sunbelt; it has merely produced gold in different, often more feverish, ways. If the American empire found its original colonies on the Western frontier, the Sunbelt cities have been its traditional outposts and provided the nodal points for its most aggressive domestic impulses.

Accordingly, these nodal points–now, sprawling Standard Metropolitan Statistical Areas–are economically "relevant." They form the centers of the "new" industries spawned by World War II, of intensively worked factory "farms," of fuels for high-technology, of shopping malls and retail emporiums. Big government, particularly the federal government, occupies every niche that has not been filled by big business and the two inevitably interlink to form big bureaucracy. Municipal autonomy has rarely been a strategic concept in the SMSAs of the Sunbelt. The earliest cities were often cavalry fortresses, not the new "Jerusalems" established by radical, often anarchistic, religious and political dissenters. Although the frontier nourished the myth of rugged individualism, its daily life and tenacious greed nourished self-interest and privatism. Not surprisingly, regional administration tends to supplant municipal administration, digesting not only neighborhoods but entire cities in the entrails of huge administrative bureaucracies. Citizenship, in turn, tends to be gauged more by the capacity to attract investment, make money, and engate in big spending than by civic activism and social reform.

The Northeastern cities are significantly different. New York, whose urban agony has made it paradigmatic for the cities of the entire region, was the most important point of entry for immigrants into the continent and their first point of contact with the realities of the "American Dream." The city achieved its elevated status not merely as a major port and financial center, but as the crucible in which the polyglot immigrants of Europe were molded into a usable labor force. American business itself accorded the city a special status, however resentfully and boorishly. Whether by virtue of high investments, political privileges or, more significantly, social reforms, the city had to be supported as the demographic and cultural placenta to Europe. More cosmopolitan than any other city in the land, it formed a lifeline to the old world with its material and intellectual riches. If a single part of the United States was the American "melting pot," it was New York City, and if America needed a space to achieve a measure of demographic and cultural homogeneity from which to draw Europe's labor and skills, it was through New York City.

The present "fiscal crisis" in New York means, quite frankly, that the city has been abandoned. Its traditional function is no longer necessary. Today, New York does not receive the bulk of its immigrants from Europe but from within the United States and its Hispanic "possessions." At a time when technology requires less muscle and more skill, New York has ceased to be an historic port of entry for needed "human resources" and has become the dumping ground for superfluous "human waste." The Statue of Liberty exhibits its backside to domestic refugees from religious and political persecution. Withs its growing proportion of blacks, Hispanics, and aged, the city has turned into an economic anachronism and a political menace. Its "minorities," who now comprise residential majorities in many parts of the city, are seen as impediments to a highly corporate, mechanized, and planned economy. Like the "masterless men" who appeared all over Europe during the decline of feudalism, these minorities have become marginal people in an era of technocratic state capitalism. From the bad conscience of the system, the city rears itself up as a specter from the past that must be exorcised. Physically it must be set adrift, abandoned to its squalor, archaisms, and leprous process of decay.

It is not a satisfactory argument to rake up the trite explanations, such as "fiscal mismanagement" and an "eroding tax base," that Washington has flung at New York to justify its neglect of the city. One could reasonably ask if Washington itself has a more sound fiscal or economic base than the cosmopolis to the north. That Washington is largely a subsidized city, indeed subsidized partly by the massive taxes it drains from New York, suggests that the viability of any city in an era of oligopoly and state controls can no longer be explained by the precepts of "free enterprise" economics. Washington is artificially sustained because it is needed as a national administrative center. To the degree that any city is a heavy recipient of direct or indirect federal funds, exorbitant revenues from oligopolistic practices, or loot drained by leisured high-income countries from exploited low-income ones, it is artifically sustained by the country as a whole. Accordingly, Washington lives on tax revenues requisitioned on a national scale; the Sunbelt cities on aeronautic and military subsidies, oil money, and real-estate hustles; the wealthy communities of Southern California on riches plundered from the poorer counties of the north and east; the Imperial Valley on artifically inflated food prices by which New Yorkers, Bostonians, and Chicagoans are bled daily. That New York has been the object of opprobrium rather than support at all levels of the federal government and the financial world is evidence not so much of its "fiscal mismanagement" but its lack of economic relevancy. Its eminence as a center of immigrant labor has waned and the immigrants it currently receives are viewed as despised social flotsam.

Perhaps no less significantly, the city has become politically dangerous. One could easily visualize that New York, which once provided the space for melding needed immigrant labor, could again be favored as a space in which "dispensable" sectors of the population could be dumped. It might seem plausible that, as "friendly fascism" oozes over the social landscape, it might leave oases in which the ethnically abused, the indigent, even the eccentrics, might find a home in the interests of social pacification. The most sinister feature of the trend toward corporate and state capitalism is that such oases are basically incompatible with a totalitarian trend, even of the "friendly" variety. The 1960s have vividly demonstrated that "affluence" does not placate the restless but awakens them. In the language of modern sociologese, improved material conditions arouse "high expectations" and ultimately a rebellious ambience throughout the country. Viewed from this standpoint, current attempts to subvert New York City's traditional reformistic policies are not without political cunning. The centralized state's growing police functions and its increasing manipulation of the economy have been followed by its growing control over local administrative authority. New York's loss of municipal self-administration to the central government could portend a far-reaching destruction of municipal institutions everywhere. Rearing up before us would be an immense political behemoth that could engulf the last administrative structures of American towns and cities.

In the Sunbelt cities, the emergence of such a behemoth already has acquired considerable reality. The tremendous weight that is given to economic expansion, to business operations, to governmentally fostered projects has served not only to promote mindless urban expansion but an appalling degree of civic passivity. The extent to which these cities have surrendered to industrial, commercial, and governmental operations is comparable only to the squalid decay of cities during the Industrial Revolution. The consequences of this surrender can be expressed as a form of municipal growth that occurs in inverse proportion to civic attrition—civic in the sense that that city once comprised a vital body politic. Homogeneity has effaced neighborhoods, regionalization has effaced municipalities, and immense enterprises, fed by the bequests of big government, have effaced the existence of a socially active citizenry. The basic concerns of the Sunbelt cities are growth, not reform; the basic concerns of its citizens are services, not social participation. Politically, the residents of the Sunbelt cities constitute a client population, bereft of citizenship and social activism by the very success of their economic growth. To the degree that meaningful politics is practiced in these cities, it is bureaucratically orchestrated by business and government.

If the great Hellenic standards of urbanism have meaning any longer for students of the city and its development, the disappearance of an active body politic, of an authentic, socially involved citizenry, is equivalent to the death of the city itself. Greek social thought viewed the city as a public arena, a realm of discourse and rational administration that presupposed a public opinion, public institutions, and a public man. In the absence of such a public, there was no polis, no citizens, no community. The Sunbelt cities have replaced public life by publicity, by a spectacular, typically

American form of "dialogue" that involves the promotion of political and economic entities. In the spectacularized world of publicity, even the classical market of free entrepreneurs is converted into oligopolistically managed shopping malls, democratic political institutions into appointed bureaucracies, and citizens into taxpayers. What remains of the city is merely its high residential density, not its urbane populace.

If the municipal success of the Sunbelt cities is marked by civic failure, the municipal failures of the older cities have been marked by a certain degree of civic success. Owing to the decline of municipal services in the older cities of the Northeast, a vacuum is developing between the traditional institutions that managed the city and the urban population itself. These institutions, in effect, have been compelled to surrender a considerable degree of their authority to the citizenry. Understaffed municipal agencies can no longer pretend to adequately meet such basic needs as sanitation, education, health, and public safety. An eerie municipal "no-man's land" is emerging between the institutional apparatus of the older cities and the people it professes to service. This no-man's land—this urban vacuum, to be more precise—is slowly being filled by the ordinary people themselves. Far more striking than New York's fiscal crisis is the public response it has evoked. Libraries, schools, even hospitals and fire houses, have been occupied by aroused citizens, a trend that is significant not because amateurs can often exhibit a technical capacity to replace the services of professionals but rather the high degree of social activism that the crisis has aroused at a grass roots level. From the seeming decline of the older cities, taxpayers are slowly being transformed into citizens, privatized districts into authentic neighborhoods, and a passive populace into an active public.

It would be naive to overstate this trend and view it as a practical solution to the crises that beleaguer the Northeastern cities. The awakening of public life in these cities will not end the erosion of their economic and fiscal bases. If the destiny of the American city is to be determined largely by its industrial and commercial "growth potential," this very criterion implies a redefinition of the city as a business enterprise, not a social and cultural space. So conceived, the city will have ceased to exist precisely because of its strictly economic preconditions and its standards of successful performance.

If the real historic basis of the city, on the other hand, is seen to be an active body politic and a spirited public life, then New York is more successful as an authentic municipality than Dallas or Houston. The evidence for this reawakening of citizen activity amidst urban decay is often compelling. For example, a convocation last year of block-association representatives by the Citizens Committee for New York City and the Federation of Citywide Block Associations yielded 1,300 activists who, according to a *New York Times* report, "debated community with the zest, and frequently the contentiousness, of an election-year political convention." The report notes that the "neighborhood activists were guided by the conviction that civic betterment starts on the block where one lives." However oppressive the problems discussed—"crime, sanitation, housing improvements, fund-raising, recycling, day care and 'fighting City Hall' "—the mood of the activists "was anything but grim. There was almost an evangelical, upbeat spirit as block leaders told of ways they had successfully dealt with safety problems or found new techniques of raising money for tree planting" (New York Times, 1976).

It matters little that the issues raised may often be trivial and inconsequential. What is far more important than the agenda of such forms is the extraparliamentary nature of the form itself and the participatory features of the association. Convocations of molecular civic groups like "block associations" that resemble a "political convention" in a normal year mark a rupture with institutionalized governmental processes. They comprise, in Martin Buber's sense, social structures as distinguished from political ones. Power acquires a public, indeed a personal, character which, to the bureaucrat, is a kind of social "vigilantism" and "anarchy" and to the participant is a "town meeting." The energy that buoys up the convocation, the anti-hierarchical character that often marks its organization, and the verve of its participants implies a renewed sense of power as distinguished from the powerlessness that constitutes the social malaise of our times.

The trivialities of the agenda should not blind us to the historic importance of municipal reawakenings at this level of action. The role of civic activism as means for far-reaching social change dates back to the American and French revolutions, and formed the basis for revolutionary change in the Paris Commune of 1871. In revolutionary America, "the nature of city government came in for heated discussion," observes Merril Jensen (1950:118-119) in a fascinating discussion of the period. Town meetings, whether legal or informal, "had been a focal point of revolutionary activity." The anti-democratic reaction that set in after the American revolution was marked by efforts to do away with town meeting governments that had spread well beyond New England and to the mid-Atlantic and Southern states. Attempts by conservative elements were made to establish a "corporate form (of municipal government) whereby the towns could be governed by mayors and councils" elected from urban wards. Judging from New Jersey, the merchants "backed incorporation consistently in the efforts to escape town meetings." Such efforts were successful not only in cities and towns of that state but also in Charleston, New Haven, and eventually even Boston. Jensen, addressing himself to the incorporated form of municipal government and restricted suffrage that replaced the more democratic assembly form of the revolutionaries of 1776 in Philadelphia, expresses a judgment that could apply to all the successful efforts in behalf of municipal incorporation following the revolution: "The counterrevolution in Philadelphia was complete."

A decade later, the French revolutionaries faced much the same problem when the *sans culottes* and *enrages* tried to affirm the power of the Parisian local popular assemblies or "sections" over the centralized Convention and Committee of Public Safety controlled by Robespierrists. Ironically, the final victory of the Convention over the sections was to cost Robespierre his life and end the influence of the Jacobins over subsequent developments. The municipal movement, indeed a rich classical heritage of the city as community that had nourished the social outlook of German idealism and later utopian socialist and anarchist theories, dropped from sight with the emergence of Marxism and its narrow "class analysis" of history. Yet it can hardly be ignored that the Paris Commune of 1871, which provided Marxism and anarchism with its earliest models of a liberated society, was precisely a revolutionary municipal movement whose goal of a "social republic" had been developed within a confederalist framework of free municipalities or "communes."

Although the older Northeastern cities of the United States hardly bear comparison with their own ancestral communities of two centuries ago, much less revolutionary Paris, it would be myopic to ignore certain fascinating similarities. The block committees of New York City are not the town meetings of Boston or the sections of Paris; they do not profess any historic goals for the most part, nor have they advanced any programmatic expression in support of major social change. But they clearly score a new advance in the demands of their participants—primarily, a claim to governance in the administration of their "blocks," a proclivity for federation, and in the best of cases, an emerging body politic. The city itself is riddled by tenants associations, ad hoc committees and councils to achieve specific neighborhood goals, a stable Neighborhood Housing Movement, and broad-spectrum organizations that propound an ideology of "neighborhood government."[2] These groups, often networks that advance a concept of decentralized self-management, however intuitive their views, stand out in refreshing relief against a decades-long history of municipal centralization and neighborhood erosion (Caro, 1974; Kotler, 1969). Even demands of "municipal liberty" are being heard in terms that are more suggestive of an earlier civic radicalism than its proponents are prepared to admit.

In a number of instances, such block and neighborhood organizations have gone beyond the proprieties of convocations, fund-raising, sanitation, public safety, and even demonstrations to take over unused or abandoned property and stake out a moral right to cooperative ownership. Apart from episodic occupations of closed libraries, schools, and a "people's" firehouse, the most important of these occupations have been neglected or unhabitable buildings. One such action, now called the "East Eleventh Street Movement" has achieved a national reputation. Initially, the Movement was a Puerto Rican neighborhood organization, one of several in the Lower Eastside of Manhattan, which formed an alliance with some young radical intellectuals to rehabilitate an abandoned tenement that had been gutted by fire. The block itself, one of the worst in the Hispanic ghetto, had become a hangout for drug addicts, car-strippers, muggers, and arsonists. Unlike most buildings which are taken over by squatters, 519 East 11 Street was a city-acquired ruin, a mere shell of a structure that was boarded up after it had been totally destroyed by fire. This building was to be totally rebuilt by co-opers, composed for the most part of Puerto Ricans and some whites, by funds acquired from a city "program" that accepts labor as equity for loans—the now famous "sweat equity program."[3] The movement's attempts to acquire the building, to fund it, to expand its activities to other abandoned structures were to become a cause célèbre that has since inspired similar efforts both in the Lower Eastside and other ghetto areas. To a certain degree, the building was taken over before "sweat equity" negotiations with the city had been completed. The city was patently reluctant to assist the co-opers and apparently yielded to strong local pressure before supplying aid. The building itself was not only rebuilt but also "retrofitted" with energy-saving devices, insulation, solar panels for preheating water, and a Jacobs wind generator for some of its electric power. An account of the conflicts between the "East Eleventh Street Movement," the city bureaucracy, and finally Consolidated Edison would comprise a sizable article in itself.

What is perhaps the most significant feature of the project is its libertarian ambience. The project was not only a fascinating structural enterprise; it was an extraordinary cooperative effort in every sense of the term. Politically, the movement was "fighting city hall"—and it did so with an awareness that it was promoting decentralized local rights over big municipal as well as big state and federal government. Economically, it was fighting the financial establishment by advancing a concept of labor—"sweat equity"—over the usual capital and monetary premise of investment. Socially, it was fighting the preeminence which bureaucracy has claimed over the community by intervening and often disrupting the maddening regulatory machinery that has so often, in itself, defeated almost every grass roots movement for structural and neighborhood rehabilitation.

All of these conflicts were conducted with a minimal degree of hierarchy and a strong emphasis on egalitarian organizational forms. Participants were encouraged to voice their views and freely assume responsibility for the building itself and the group's conflicts with municipal agencies and utilities. This organizational form has been preserved after the rebuilding of 519 East 11 Street was completed. The entire block was—and, in part, remains—involved in varying degrees with the group's activities and its efforts to reclaim other buildings in the area. Many participants have acquired a heightened sense of social awareness as a result of their own efforts to achieve a degree of "municipal liberty," if only for their own physical space and nearby blocks. Activists who remain involved with the larger aspects of the project—its explosive political, social, and economic implications—have a radical consciousness of their goals. What began as a desperate effort of housing co-opers to rescue their neighborhood, in effect, has become a social movement.

Such movements, in some cases involving "illegal" seizures of abandoned buildings, are growing in number in New York and other older cities. Although they have not always exhibited the staying power and libertarian ambience of the "East Eleventh Street Movement," they must be seen in terms of the context they have themselves created. "Municipal liberty" in the older cities, to be sure, does not mean the "liberty, equality, and fraternity" which the more radical Parisian sections tried to foster; nor does it have the mobilizing and solidarizing qualities of the more radical American town meetings. The projects that can be related to this new civic trend—be they housing co-opers, "sweat equity" programs, block committees, tenants groups, neighborhood "alliances," or cooperative day-care, educational, cultural, and even food projects—vary enormously in their longevity, stability, social consciousness, and scope. In some cases, they are blatantly elitist and civically exclusionary. To a large extent, they form a constellation of new subcultures that have evolved from the broader countercultural movement of the 1960s, a constellation that has been greatly modified by ethnic disparities, urban disarray, a broad disengagement of municipal government from its own constituency, an emerging "free space" for popular, often libertarian, civic entities, and the civic bases for a new body politic.

But a living trend they remain—and the most important trend to emerge in American cities today. In contrast to the bureaucratically managed and municipally regimented Sunbelt cities, they represent a largely regional development. The very fact that they have been fueled by urban decay conceals their significance as the most significant trend in generations to reclaim the city as the public space for an authentic citizenry. If they are not a "vision" of the future, they may well be one of its harbingers. Certainly they are one of the most exciting links American cities have yet produced between the urban past and the urban future—a new "treasure," as Arendt might have put it, in the development of human community and the human spirit.

A vision of the urban future—if it is to be conceived as a city and not a sprawling agglomeration of man-made structures—is haunted by the past. The trite assumption that we cannot "return to the past" can become an excuse for ignorance of that very past or an unconditional renunciation of what we can learn from it. To the serious student of urban life, the most fascinating point of departure for relating past to future is the Hellanic polis. That we live in a world of nation-states and multinational corporations is no excuse to continue to do so. The urban future must be viewed from a standpoint that may sharply contradict the immediate future of our present SMSAs, a future that seems to consist of more business, more structural as well as economic growth, and more centralization. That future must be above all a new conception of the city that fulfills our most advanced concepts of humanity's protentialities: freedom and self-consciousness, the two terms that form the historic message of Western civilization.

Self-consciousness, at the very least, implies a new self: a self that can be conscious. Consciousness, certainly in the fulness of its truth, presupposes an environment in which the individual can conceptually grasp the conditions that influence his or her life and exercise control over them. Indeed, insofar as an individual lacks these dual elements of consciousness, he or she is neither free nor fully human in the self-actualized sense of the term. Denied intellectual and institutional access to the economic resources that sustain us, to the culture that nourishes our mental and spiritual growth, and to the social forms that frame our behavior as civilized beings, we are not only denied our freedom and our ability to function rationally but our very selfhood. The great cultural critics of society have voiced this conclusion for centuries. This conclusion has even more relevance today—an era of social decay that seems almost cosmic in its scale—than at anytime in the past.

In terms of the city, such a conclusion means that a vision of the urban future can be regarded as rational and humanly viable only insofar as the city lends itself to individual comprehension—notably, that it is an entity that can be understood by the individual and modified by individual action. That the city whose population "can be taken in at a single view"—that is, scaled physically and numerically to human dimensions—remained essential to the Hellenic ideal of the polis is merely another way of saying that a city without a citizenry, an active body politic, is not a city, indeed unworthy of anything but barbarians (Aristotle, 1943:VII, 5:25). Human scale is a necessary condition for human self-fulfillment and social fulfillment. A humanistic vision of the future city must rest on the premise that the authentic city is comprehensible to its citizens or else they will cease to be citizens and public life itself will disappear. A vision of the urban future is thus meaningless if it does not include from its very outset the decentralization of the great SMSAs, the restoration of city life as a comprehensible form of public life.

Still another vision of the future must include the recovery of face-to-face forms of civic management—a selfhood that is formed by self-management in assemblies, committees, and councils. We can never "outgrow" the Hellenic ecclesia or the American town meeting without debasing the word "growth" to mean mere change rather than development. The existence of an authentic public presupposes the most direct system of communication we can possibly achieve, notably, face-to-face communication. Again, another of Aristotle's (1943: VII, 5:15) caveats is appropriate here: "in order to decide questions of justice and in order to distribute the offices according to merit it is necessary for the citizens to know each other's personal characters, since where this does not happen to be the case the business of electing officials and trying law-suits is bound to go badly; haphazard decision is unjust in both matters, and this must obviously prevail in an excessively numerous community." It need hardly be emphasized that Aristotle would have been appalled as much by the telecommunications of a "global village" as he would have been by the very concept of the world as a huge city or village. Human scale thus means human contact, not economic, cultural, and institutional comprehensibility alone. Not only should the things, forms, and organizations that make up a community be comprehensible to the citizen, but also the very individuals—their "personal characters"—who form the citizen body. The term "citizen body," in this sense, assumes more than an institutional concept; it takes on a physical, existential, sensory, indeed protoplasmic, quality.

Thus far, I have been careful to stress the conditions that foster public life rather than the things that make for the "good life" materially. Decentralization and human scale have been emphasized as the bases for a new civic arena. Whether they are more "efficient" systems of social organization or more "ecological" types of association, as some writers have argued, has not been emphasized.[4] That a city, landscaped into the countryside, will promote a new land ethic and afford its citizens greater access to nature—perhaps even restore the urbanized farmer so prized by the Athenian polis and republican Rome—adds to the case for a more rounded body politic in a more rounded vocational and physical environment. But ultimately it is the very need for a reactivated citizenry that must be stressed over efficiency, ecological awareness, and vocational roundedness. Without that citizenry we now face the loss not only of our cities, but of civilization itself.

Finally, the recovery of a body politic and a civic community can scarcely be imagined without the communitarian sharing of the means of life—the material as well as social communizing that authentic community presuppose. In a technological world where the means of production are too powerful to be deployed any longer for means of domination, it is doubtful if society, much less the city, can survive a privately owned economy riddled by self-interest and an insatiable need for growth. More than the "good life," materially speaking, is involved in a communitarian system of production and distribution; the very existence of a coherent community interest is now at issue. Here, too, Hellenic culture has much to teach us about the future. Private interest can not be so dominant a motive in social relationships that it subverts the public interest. If private property once formed an underpinning of individualism in the corporatized cities of the past such as the guild-directed medieval towns, today, in the "free market" of giant oligopolies, it has become the underpinning of naked egotism, indeed, the institutionalized expression of a social behavior of the most ruthless kind. If the city is to become a public body of active citizens, it must extend the public interest to the material as well as institutional and cultural elements of civic life.

Here we can part company with the Hellenic outlook and view the future as more than a recovery of the past. Modern technology—"hard," "soft," "appropriate," or as I would prefer to call it, liberatory—has finally made it possible for use to eliminate the fears which stalked Aristotle: "an overpopulous polis [of] foreigners and metics [who] will readily acquire the rights of citizens" (1943: VII, 5:20). To these potential "upstarts" one might also add slaves and women. The leisure or *scholē*—the freedom from labor—that made it possible for Athenian citizens to devote their time to public life is no longer a birthright conferred by slavery on an ethnic elite but one conferred by technology on humanity as a whole. That we may feel free to reject that birthright for a "simpler," "labor intensive" way of life is a historic privilege that itself is conferred by the very existence of technology. Although a "global village" created by telecommunication would be an abominable negation of the city as a citizen body, "global citizenship" in clearly defined cities would constitute its highest actualization—the civic socialization of parochial folk into a universal humanity.

This vision of the urban future must now stand as it is—vague and broad, but hopeful. Any additions or details would be utopian in the worse sense of the word. They would form a "blueprint" that seeks to design without discussion and impose without consent. A libertarian vision should be a venture in speculative participation. Half-finished ideas should be preferred deliberately, not because finished ones are difficult to formulate but rather because completeness to the point of details would subvert dialogue—and it is dialogue itself that is essential to civic relations, just as it is *logos* that forms the basis of society.

NOTES

1. The ambiguity of the tendency is evident in the writings of Marx and Engels. Despite Engels's critical thrust in his well-known pamphlet "The Housing Question," he clearly shared Marx's view that the bourgeois city marked a distinct advance over rural "parochialism."

2. Notably organizations such as the Alliance for Neighborhood Government in the United States and the Montreal Citizens' Movement (MCM) in Canada. The MCM, which already holds a considerable number of seats in Montreal's city council, has advanced the most radical program of all. "Nous devons instaurer notre propre democratie afin de realiser notre plan de reorganisation de la societe," it declares in its latest program. And further: "Le conseil de quartier (which the MCM seeks to substitute for the existing "districts electoraux") ne devra donc jamais devenir un autre palier de gouvernment a interieur de la societe capitaliste" (Montreal Citizens' Movement, 1976).

3. There is, in fact, no offical "sweat equity program" in New York City. The "program" is the legal and funding nexus which youthful activists on the East 11 Street project and "U-Hab," a New York homesteading group, created when early attempts were made to rebuild abandoned structures in the city. For the most recent survey of "sweat equity" projects in New York, see the Third Annual Progress Report of the Urban Homesteading Assistance Board.

4. Milton Kotler (1975), for example, has emphasized the efficiency of decentralization and F.S. Schumacher (1973), its capacity to promote ecological awareness. In the latter case, I must share some responsiblity for this emphasis in as much as Dr. Schumacher, quoting me by earlier pseudonym, Lewis Herber, accepts my assertion that "reconciliation of man with the natural world is no longer merely desirable, it has become necessary" (p. 107).

REFERENCES

ARENDT, H. (1954). Between past and future. New York: Viking.

ARISTOTLE (1943). Politics (B. Jowett, trans.). New York: Modern Library.

BOOKCHIN, M. (1973). The limits of the city. New York: Harper and Row.

——— (forthcoming). Urbanization without cities. San Francisco: Sierra Club.

CARO, R.A. (1974). The power broker. New York: Knopf.

DOXIADES, C.A., and DOUGLAS, T.B. (1965). The new world of urban man. Philadelphia: University of Pennsylvania Press.

JENSEN, M. (1950). The new nation: A history of the United States during the confederation, 1781-1789. New York: Knopf.

KOTLER, M. (1969). Neighborhood government. New York: Bobbs-Merrill.

--- (1975). "Neighborhood government." Liberation, 19(8 and 9):119-125.

Le CORBUSIER (1971). The city of to-morrow. Cambridge: M.I.T. Press.

McLUHAN, M., and FIORE, Q. (1968). War and peace in the global village. New York: Bantam.

New York Times (1976). March 26.

SCHUMACHER, F.S. (1973). Small is beautiful. London: Blond and Briggs.

WRIGHT, F.L. (1964). "The city as a machine." Pp. 91-94 in C.E. Elias, Jr. et al. (eds.), Metropolis: Values in conflict. Belmont, Calif.: Wadsworth.

12

People, Profit, and the Rise of the Sunbelt Cities

DAVID C. PERRY
ALFRED J. WATKINS

□ THE DOMINANT CHARACTERIZATIONS of the present state of uneven urban development mask a telling contradictory view of urban America. For the urban poor, the "rise of the Sunbelt" and the "decline of the Northeast" are meaningless descriptions. Their poverty is not a regional issue. But the vast outpourings of many journalists, academics, and politicians seem to indicate otherwise: low income in the Northeast represents "poverty" while low income in the Sunbelt represents the advantage of cheap labor and is part of a "good business climate."

At the heart of this disagreement are competing definitions over the role the city should play in a society. For the established analysts, the primary function of the American city is its productivity function—the generation of profit. The prime measures of urban health or dynamism are economic and these analysts discuss the growth of cities in terms of their success or failure as centers of profit; sorting them out along continuums of central city and suburb and Northern and Sunbelt. For the poor, the function of the city is only secondarily associated with its productivity function. The city, whether it is economically profitable or not, should be first and foremost a center of social well-being, providing a community which is at once materially and socially renewing.

These divergent conceptions of the fundamental role of the city in America form the basis of this essay. In the two sections which immediately follow we briefly discuss the emergence of the city as first and foremost a center of profit in America. In the third section we review three waves of migration to American cities testing the notion that the conditions of urban poverty are not irrevocably altered by the economic rise or decline of a city. What is altered during such periods of change in urban productivity is the ability of the capitalist society to economically utilize and otherwise socially control the poor. In the last two sections of this essay we argue that continued adherence to the dominant role prescribed for the American city will have no significant impact on the existence of a permanent urban underclass. Unless we begin the socially necessary task of recasting the role of the city in our society so that it is primarily a center of people and not profit, we will never fully understand much less witness an end to the domestic tensions fostered by the contradictions enamating from competing definitions of the purpose of our urban society.

THE URBAN FUNCTION IN AMERICA

Before proceeding to the empirical tasks set forth in the introduction, a short discussion of our concept of the prevailing role assigned to the American city is in order. It is axiomatic to say that every society, in order to survive, must be productive—that is, it must generate material surplus. What differentiates societies are the purposes to which this surplus is directed. At the most basic level, societal productivity can be either accumulative or redistributive. If it is accumulative, it allows for an uneven distribution of surplus to its citizens, each individual rightfully accruing his share of the surplus because of his owned share of the means of production. On the other hand, surplus used redistributively accrues to citizens evenly, based on their political right to share equally in the productive fruits of society. Both these pure political-economic states can and do use surplus *replacively* to guard against disaster or refurbish the productive infrastructure, and *wastefully*, misspending surplus on luxuries which have no redeeming productive value. Finally, we would argue both political economies can *misallocate* surplus. That is, capitalist economies can err and use surplus in a nonreplacive and overly redistributive way and socialist economies have the untoward potential to use capital in a nonreplacive and overly accumulative manner.

Obvioulsy this simple political-economic conceptualization of societies is an "ideal type." There are no pure politically individualistic, economically accumulative, nation states and no pure politically collective, economically redistributive states. The real world is characterized by the thin line that nations dance between preserving their predominant political economic nature and committing irreversible errors in the misallocation of their productivity. The government is the institution called upon to "choreograph" such a "dance" and such a role is usually called the social control function. Ira Katznelson captures the essence of the social control function of the American state when he declares: "The state's function of social control consists in managing the consequences of making capitalism work and can best be understood as an attempt to manage but not overcome the contradictions of the capitalist system" (1976:220). Interpreting this description for our purposes, we can say that if the state fails in such management or choreography of the tension between the redistributive demands of the "nonowners," or the poor, and the predominant political economic order, then misallocation of surplus (in the name of rising public services) or the implementation of a police state follow. In either case, the predominant order; the legitimacy of the society has been threatened by the presence of militant or at least economically "unmeltable" citizens. The presence of a socially uncontrollable and economically unprofitable class of people represents a crisis in the legitimacy of the capitalist society itself.

In this context, it can thus be argued that studying the migration of various groups of poor people to the American city gives us a clear opportunity to place in larger perspective what is meant in the prevailing discourse by the "decline" of the urban Northeast and the "rise" of the Sunbelt cities. Indeed it is our thesis that urban poverty has become a symptom of decline in the Northeast because it is no longer viewed as either economically or politically manageable. Northeastern cities are constantly "misallocating" larger and larger shares of public surplus to meet the increased demands of economically "illegitimate" people. A contemporary way of describing this misallocation is to talk of an urban "fiscal crisis." Conversely, the "rise" of the Sunbelt cities represents a "good business climate" because poverty implies that people are willing to work for low wages and because public service burdens have not yet reached the stage where they could be defined as a misallocation of the productive surplus.

However, the veracity of this version of regional uneven development is not found in the simple model presented here, it is found in the experience of successive waves of urban migrants as they came to our nation's cities. We now turn or rather, in this paper, "dash" through history in an attempt to provide a first approximation of our thesis on the evolving role of the city in America.

THE CITY, THE PROFIT FUNCTION, AND THE EARLY IMMIGRANTS

Few nations have grown to urban status in quite the way of the United States. The American city did not evolve so much as it exploded. In less than 160 years we have changed from a time when less than 10% of our population lived in urban places to a time when three-quarters of us are urbanites. Relative to other advanced nations, our cities are not the result of centuries of cultural growth; they are the inventions of decades—manufactured as if overnight, out of whole cloth (Bookchin, 1973:90-93). As a result, even the oldest of our cities have no long standing cultural history derived from a singular people; rather they have served as sorting out and resocialization centers for successive waves of racially and culturally diverse outsiders. From the beginning, our cities fueled the dreams of materially oppressed immigrants; they were centers of individual gain, with "streets paved with gold." The fulfillment of such a dream was more than enough to force the immigrants into a Faustian contract with the New World; to exchange the "soul" of their homeland for the "gold" of the open land and later the cities of America. In short, the history of the American city—Yankee and Sunbelt—is the history of the rise of the *capitalist city* (the center of material profit) on the initially expansive urbanless *tabula rasa* of the New World.

This conception of the economic function in the New World was not central to the immediate dream or assumptions the early settlers had about their lot in life. For the first 200 years it appears that the New World settlers were basically hopeful of being "less poor," but poor just the same. Their heritage of Old World poverty was far from an entrepreneurial one. Robert Bremner (1956:3) points out that "during the first two centuries of the country's development most Americans took it for granted that the majority of men would always be poor. Poverty was the state from which thousands of emigrants fled when they embarked in hope or despair on the difficult journey to the New World, in the form of hardship, privation and suffering it was the lot, not of the first settlers on the alien coast, but of generations of pioneers on successive inland frontiers."

However, the growth of our first cities into mercantile centers and later on the transformation of others into industrial centers of profit brought about unprecedented wealth for some. This overall process is depicted by historian San Bass Warner as the process of "privatism." He describes the requirements of life in the American city as follows:

> Under the American tradition, the first purpose of the citizen is the private search for wealth; the goal of a city is to be a community of private money-makers. Once the scope of many city dwellers' search for wealth exceeded the bounds of their municipality, the American city ceased to be an effective community. Ever afterwards it lacked the desire, the power, the wealth, and the talent necessary to create a humane environment for all citizens. From that first moment of bigness, from the mid-nineteenth century onward the successes and failures of American cities have depended upon the unplanned outcomes of the private market's demand for workers, its capacity for dividing the land, building houses, stores and factories, and its needs for public services have determined the shape and quality of America's big cities. What the private market could do well American cities have done well, what the private market did badly, or neglected, our cities have been unable to overcome. [1968:x]

Thus, the city in America emerged as a marketplace for competitive material productivity: a profit place not a social place. The social well-being of its immigrating citizens could be accomplished through their economic renewal (James, 1972:22-30). While it became rationalized economically, the city remained a socially disintegrated community, sorting out successive waves of immigrants within the criteria of profitable productivity. Beyond this, immigrants were left relatively alone with the cultural heritage of their immediate past to salvage their own social community (or, as we have come to call it, ghetto).

PROFIT, POVERTY, AND THREE WAVES OF URBAN MIGRATION

Our thesis in this section, quite boldly, is that for most of our history the place of the urban poor has been comfortably "hidden" under the "rock" of urban profit. America becomes concerned with its urban poor when their conditions can no longer be hidden; when they pose a barrier to economic growth and a threat to the social order, thus exacerbating a developing crisis in the legitimacy of the city as a center of profit. More precisely, we will argue that the poor emerge as burdens when they can no longer be used productively (as a supply of cheap labor or rental income) or controlled (as they come to represent a threat to the social order through health hazards, riots, racial tensions, increased wage demands, anarchistic acts of violence, or organized political movements).

THE EUROPEAN IMMIGRANT AND THE NEW CITY OF PROFIT

Early in the 19th century, as Americans began to experience the first dramatic blush of urban-based affluence, they also began to observe new levels of urban poverty (Greeley, 1972). The close proximity of rich and poor in the city set out in dramatic relief growing social disparities arising from rapid immigration, industrialization, and urbanism. Such a disparitous condition did not scare people away; in fact it had the opposite result. It removed the vision that the state of man was inherently one of poverty; replacing this notion was the dream that America was the place of opportunity, of the *economic* "Horatio Alger," of the *political* "Lincoln Ideal" (Ginger, 1973). The emigrant from the Old World or the farm could go to the city and literally walk along streets "paved with gold." The American city was the place for living the liberal-capitalist dream, where ostensibly one was at last free to contract one's labor, and where one was free to accumulate wealth, through ownership and even profit making.

Evidence of such dreams lies in the demographics of 19th century America which exhibited one of the most dramatic immigration and urbanization rates in the history of the Western world. The national population sorted out in two different ways: first, through an explosion and expansion across the breadth of the continent spurred by an entrepreneurial process of frontier development (Wade, 1959;

Chudacoff, 1975); and, second, through a veritable population implosion in our industrializing cities. This second demographic pattern was a boon to the capitalist city in two major ways. It provided an apparently limitless supply of cheap labor, thus ensuring low wages and high profits in the emerging manufacturing sector. Second, the very presence of literally millions of new people in our cities, made for an almost inexhaustible source of residential demand. "Proper" land development in the city–filling every available foot of a slum lot with cheap housing and filling every available foot of a house with humans–represented a highly profitable practice of land development. Perhaps tenement slum development did not offer quite the dramatic level of profit to be found in frontier land speculation, but the returns to investment were more immediate and the speculative risks were negligible.

By the 1840s tens of thousands of immigrants, many from Ireland, were streaming into Northeastern American cities. Numerous authors have given us chilling descriptions of the housing conditions found there. In New York tenements Charles Dickens saw attics filled with startled wretched creatures who would crawl from infested nests and corners "as if the judgment hour were at hand and every obscene grave was giving up the dead" (1972:137-138). Jane Addams (1910) witnessed the impact of similar conditions on tenement dwellers in Chicago, and William Ellery Channing reported that many tenements in Boston consisted of "cellars and rooms which could not be ventilated, which want for benefits of light, free air and pure water, and the means of removing filth" (Bremner, 1956:5).

However appalling the conditions of slum dwellers during the middle of the 19th century, these were rarely viewed as evidence of an "urban crisis." On the contrary these people and the very oppression of their existence represented a true economic resource in the American process of urban dynamism. Their very numbers worked against them, providing the city with a limitless pool of labor and reducing the bargaining power of the unskilled workman with industrial owners to near zero. Because the male worker was further forced to compete not only with other incoming immigrant males but with women and children (Bremner, 1956:4), his wages often never exceeded $5.00 a week (Gambino, 1975:92).

Italians, largely farmers before their emigration, moved to the cities of the United States in great numbers after 1880. Upon their arrival, they were employed in the "hard, dirty, usually menial labor . . . left behind by the North and West Europeans who had moved up the socioeconomic scale" (Feagin, forthcoming). Their work conditions, like those before them, were also abominable:

> The Italian immigrant may be maimed and killed in his industrial occupation without a cry and without indemnity. He may die from the "bends" working in the caisons under the river, without protest, he can be slowly asphyxiated in crowded tenements, smothered in dangerous trades or occupations (which only the ignorant immigrant pursues, not the native American); he can contract tuberculosis in unsanitary factories and sweatshops. [Feagin, forthcoming, Chapter 6]

In fact, tuberculosis, a rarity in Italy, was a major source of death in the urban tenements of our cities' "little Italies" (Feagin, forthcoming).

Moreover, by the late 1880s, the history of the Italian in-migration shows that labor conditions for immigrants and tenement dwellers had not been substantially altered, even with a growing union movement (Grob, 1969). In 1895, the day scale of work for manufacturing laborers, in the "gold paved" streets of New York, was listed in a newspaper advertisement as follows (Gambino, 1975:77):

> Common labor, white, $1.30 to $1.50;
>
> Common labor, colored, $1.25 to $1.40;
>
> Common labor, Italian, $1.15 to $1.25.

In short, as the 1910 Industrial Commission on Immigration put it: "Where an Irishman or a German demands meat, an Italian will work upon stale bread and beer, and, although his physical efficiency is not as great, he works for so much less that it is *profitable* to employ him" (Gambino, 1975:87, emphasis added).

At about the same time, the front sections of the newspapers of New York heralded the emergence of the city as the rightful new center of American life, growth, and prosperity (McCabe, 1882). Historian Roy Lubove describes the end of the 19th century as the time when cities "shattered the Jeffersonian-Jacksonian vision of the yeoman, agrarian republic. The economic vitality of the new era centered in the factory, not the farm. The city, rather than the small town, became the *undisputed symbol of America's productive energies,* cultural and intellectual attainments, economic and social opportunities" (1962:40). The measures of the city's success as a center of profit hid its failures as a center of social well-being.

There were those who argued for social reform, and against the impact of the excesses of profit making on poor city dwellers. But such critics were few and far between and their propositions for reform, as well as their critiques, rarely made them advocates for redistributively reshaping the city's character in the name of the people. In fact, one group of advocates for slum renewal put the blame for depressed wages and slum conditions at the feet of the immigrant. This early form of the social control strategy of "blaming the victim" for urban poverty, argued that our cities could be returned to a wholesome state when we stopped admitting overwhelming numbers of inferior aliens to our cities (Morse, 1835; Greeley, 1864).

A more sympathetic version of such ethnocentric reform argued that even though the immigrants were marginal earthlings, to be treated by all normal souls with understandable repugnance, their very presence and life-styles could not be ignored (Bremner, 1956:6; Hartley, 1842:18; Dickens, 1972). The slum, and the disease it spawned, was a potential threat to *all* city dwellers. And the most *successful* early reformers quickly couched arguments for changing the outrageous living conditions of slum dwellers from Boston to Chicago to St. Louis in terms, not of helping the poor, but of controlling their presence in order to protect the social order. A three part platform of 19th century urban reform emerged looking something like this: (1) establish standards for sanitation in the slums to guard the *public* health; (2) rectify living conditions in the slum as a guard against the *spread of the "dangerous classes";* and (3) remove slum conditions in areas *unprofitably close* to the business districts of the city (Brace, 1872; Hartley, 1842; Griscom, 1845). The reform for both public and private sectors now took the form of regulating the victim rather then simply blaming him (Piven and Cloward, 1971; Trattner, 1974).

LuBove (1962:48) points out that the first major public hygiene reforms occurred in New York City with the creation of the Metropolitan Board of Health. The board was created not out of a concern for the living conditions of the poor, but after a cholera epidemic threatened the health of the entire city. The first public standards affecting the structure and design of tenements were not directed at the inhuman nature of poor housing conditions; rather they were offered as a way of mollifying the "dangerous" residents who had rioted over military draft conscription policies in 1863. Indeed, it appears that "terror had succeeded where reason, enlightened self-interest and pleas of humanitarians had failed" (LuBove, 1962:23).

The reforms for greater welfare were further modified by the opportunities of profit realized in the tenement districts. Besides the bounty of cheap labor, the urban slums of the 19th century also represented a growing place for rental profit. Thus a successful reform soon became one that balanced the rights of people to decent housing against the rights of the urban entrepreneur.

On the surface, local legislation dictating the construction of tenements with light shafts, air vents, fire escapes, larger water closets, privies, and other amenities appeared to be quite sensible when compared with the conditions described by Dickens, Hartley, and others.

However, even these basic changes were not often provided. Attempts were made to include these amenities in the construction of "model tenements," which were designed to reduce overcrowding and profit gouging. To accomplish the latter, it was argued that the profits generated by these structures should be limited to 4% or 5% rather than the usual 15% to 18%. Few such structures were ever built (LuBove, 1962:25-33, 86-115). Another reform-based architectural structure was very popular—the "dumb-bell" shaped tenement. It was the winning entrant in a nationwide contest designed to develop the most "healthy and morally" spacious tenement that could be constructed on a city lot measuring 25 by 100 feet. The purpose of the structure was to reconcile the tenant's welfare and the investor's profit. Overnight it became a characteristic type of tenement in working class New York and other cities. It usually stood around six stories high, with 14 rooms to a floor and it housed two dozen families.

Such forms of social reform obviously did little to benefit the slum but they do not appear to have hurt the tenement landlords. Indeed, the "dumb-bell" and other such reforms appear to have been good business. By 1890, 35,000 of New York's 81,000 dwellings were tenements ("reformed" or otherwise). Of the 1.5 million people living in New York, 1 million of them lived in these slum dwellings. By 1916, the number of New York tenements had increased by 5,000, yet the number of tenement dwellers had increased to an incredible 2,082,000 (LuBove, 1962). While the absolute magnitude of this trend was not as great in other industrial cities, they did exhibit similarly dramatic practices of urban profit at the expense of an exploited underclass (Chudacoff, 1975).

Again it is important to point out that these cities were viewed as the vital new social and economic centers of America (LuBove, 1962). Their poor were not barriers to growth; in fact, they were in many ways just the opposite. As sources of cheap labor and as rental markets, they were a source of "locational advantage" for the capitalist city.

Urban poverty was certainly not defined as an "urban crisis." From this time forward the depressed conditions of urban life would only come to be viewed as an urban crisis when they posed a threat to the economic dynamism (profitability) of the city. Public and private sector reforms would be instigated primarily to control such threats, thus protecting the economic vitality of the "private city."

The preceding historical discussion can now be used to frame two more recent patterns of migration to American cities: the emigration of rural Southern blacks to the cities of the North and West; and the most recent movement of Northern urbanites to the cities of the Sunbelt.

THE BLACK MIGRANT AND THE CHANGING NATURE OF URBAN PROFIT

There are significant differences as well as significant parallels between the black American's arrival in industrial cities and the earlier entry of European migrants. Like many of its ethnic predecessors, the black population which pushed north after 1940 was a basically *rural* and *unskilled* group. Blacks were also a marginal group; they broght with them a heritage, race, and presence that was highly unacceptable to white America. But their racial marginality represented a form of social identity which had not accompanied the European immigrants. Blacks were not aliens emigrating from a foreign country. On the contrary, they were American residents of long-standing second-class status: a status which had evolved historically as predominantly Southern, rural, and poor.

Between 1940 and 1966, the migration of 4.7 million of these people to the industrial North and the West precipitated a profound change in their residential and geographic character. They were reduced by technological changes in agriculture to a position of uselessness and, driven by "imminent starvation and eviction" (Cloward and Piven, 1974:x), they made their way to the cities. Thus, where 80% of our black population lived in the South in 1940, today only 53% reside there. And where, in 1940, 32% of all employed blacks worked on farms, today only 2% are farm workers. The result of these shifts is that a higher proportion of blacks than whites now live in metropolitan areas (Farley, 1977:189). In less than four decades black Americans suffered a dramatic dislocation that had a profound effect on their previous social and economic status.

While blacks did not move directly from farms to the major cities, once in urban areas, they were forced by discrimination to gather in ghetto communities. In many cases these ghettos were the slum-worn neighborhoods that had acted as staging areas for the advancement of past marginal groups. But while these areas would serve as places of cheap housing, communal centers of social reinforcement and friendship, and supply centers of public services as they had for past groups, they would not be the economic launching pads they had been previously. On the contrary, these neighborhoods have become final repositories for vast numbers who were pushed by poverty from the farm to the city, only to be "pushed around" again.

It is too simple to conclude that if blacks had come 50 years earlier that such slums would have been launching pads for their socioeconomic integration. Their present day economic position is tied to a variety of features which differentiate their arrival in the industrial city from that of other immigrants. First, their position in American society as a racially marginal group is one of long standing historical duration. The vestiges of racism remain fundamentally intact to this day. As the agricultural economy of the United States diversified and its productivity increased with technological change, blacks lost rather than gained ground. On the other hand, the urban economies to which they moved no longer needed the large numbers of unskilled labor that had originally primed and fueled their early growth. The Northeastern industrial economies were just not growing very rapidly anymore and where there was growth, blacks had to compete with *white* unskilled labor for jobs (Grob, 1969).

As in their days in the rural South, changes in the urban economy have meant growing sectors of black unemployment and poverty. In short, a racial group, which outlived the economic value society had assigned to it after the Civil War, found the same history of socioeconomic uselessness assigned to it in the city.

However, one component had changed; the economic vitality of the cities themselves was now, in some cases, becoming somewhat suspect. As the power of the accumulation function waned, not only as an avenue of profit, but also as a manipulator of the cheap labor and slum conditions of the poor, the legitimacy of the role of the cities as centers of this accumulation was weakened. In the Northeastern urban areas, characterized by increasing wages, job benefits, and union organizing, rising material expectations among consumers and growing civil rights legislation, the black urbanite was justifiably incensed by the failures of the declining city. The riots of the 1960s are ample evidence not only of black rage but also of the growing failure of the city's accumulative capacity to utilize and thus manage poverty profitably. Blacks were not needed as cheap labor and without jobs they represented limited consumption power. Without such consumption ability, they did not represent a new profit market of immigrant renters. *The Northeastern city was as useless to them as they were to it.*

As such, by the 1960s the presence of black immigrants represented more than an economic liability; they also represented social control burdens in urban centers already feeling the first major ripples of economic stagnation and decline. The explosion of these cities into riot torn war zones displayed in dramatic relief the failing legitimacy of these cities in the eyes of urban black America. It is important to note that the major riots of this era did not break out in Sunbelt cities like Houston, Dallas, Austin, Miami, or Phoenix. For the most part, the major race riots occurred in economically declining urban centers of the North where, as our analysis suggests, the poor could no longer be profitably absorbed or efficiently managed by the economic and political sectors of the city.

Finally, the national political climate of the nation between the end of World War II and the beginning of the 1970s was substantially different than it was during previous eras of urban in-migration. The most dramatic difference was the centralization of many urban social services at the federal level. Again, race had much to do with this. While urban blacks had come to represent a significant electoral force in our nation's cities, few had become integrated into the political machinery of city politics. They did not receive, in the form of social services, public jobs, and political power, the patronage due them for their vote. Frances Piven argues that this time-honored form of integration of the urban underclass into city society did not occur because, although national democratic presidents such as Kennedy and Johnson owed much of their strength to the Northern urban black vote, local city politicians derived their structural power from old, predominately white, political ties (Cloward and Piven, 1974:274-275). In part, in an attempt to break the blockage of black integration through patronage, the Kennedy-Johnson administrations launched a full phalanx of "categoric" urban aid programs, many of which were specially targeted toward the politically vote-rich, yet slum-infested, wastelands of our Northeastern industrial cities.

With the advent of the riots in cities losing both economic vitality and social control over the poor, the federal government, like social reformers of bygone eras, responded not simply out of political patronage, but also out of fear. The "war on poverty" escalated in cities of declining profit to the point that by 1969, it appeared that the most profit to be made in some of these cities was in the "inefficiency-laden" public service deliv ry systems themselves. If one reads the legacy of these programs, setting aside for the moment that they were born primarily out of political pragmatism rather than a redistributive spirit of social well-being, it is apparent that their failure has been measured more by how costly they were than by how little they changed the lives of the urban poor. In short, the "war on poverty" has been reduced to a "war against welfare cheaters." A rather sophisticated new welfare-labor-housing political rhetoric of blaming the "able-bodied" victim has emerged as the urban "fiscal crisis" of the 1970s, replacing the urban social crisis of the late 1960s. The political strategy of spending the economically declining Northeastern cities out of their 1960s malaise failed because the private sector was deserting the Northeast at a time when the social welfare dollars of federal and state governments were not enough to trigger social renewal in a milieu of both declining profit and declining urban legitimacy. Within the frame of this essay, then, continuation or expansion of the social reforms of the 1960s in this climate represented for the dominant order a serious *mis*allocation of declining surplus; indeed it came to be deemed one cause of our Northeastern "fiscal crisis."

Hence, 130 years after Charles Dickens walked the tenement streets of Manhattan, another journalist, Jimmy Breslin, walked the Brownsville section of Brooklyn and wrote not of teeming tenements but of a semi-deserted, physically decimated, and rat-infested community of economically vanquished people:

> At night, the streets of Brownsville are like a well bombed target. On each block there are half-demolished buildings. Their corroded insides of staircases and broken walls and sagging floors are outlined by the car headlights. . . . The people in these lone apartments must keep somebody awake all night, because the kids from the neighborhood come into these buildings and set fires in the empty apartments. . . . Here and there in the ruins of Brownsville there are neatly painted wooden signs proclaiming that a housing project will be erected on the spot. Under the proclamation is the name of the politician, and of the various urban experts, in charge of the housing program. These signs have been standing in Brownsville for many, many months. Just as the same signs have been standing for the same months and months on Roosevelt Avenue in Chicago and Twelfth Avenue in Detroit and Joseph Avenue in Rochester. The story of a city in this nation in the 1960s is a sign with a politician's name on it, and only the name on the sign changes. [Breslin, 1971:233]

Brownsville represents more than poverty; its declining condition represents a society pulling back on the reins of social reform. Apparently there is no redeeming productive value (Baer, 1976; Starr, 1976) to be found in the ghettos anymore and the shift in the social control strategy of the state is as clear as the fading paint on the state program signs.

The contradiction represented by Brownsville and scores of other Northeastern inner cities is that millions of their residents are useless to the city—their needs represent a strain on the accumulation function with no returns to urban profit. Conversely, the historic racial and economic illegitimacy of blacks is only matched by the political and economic illegitimacy of the city to urban blacks. In 1976 it is estimated there were 30,000 burned residences in the South Bronx section of New York City alone. Nationally, thousands of assaults reportedly occurred against inner city school teachers and welfare workers, and the welfare system is in shambles. Daily, guerilla warfare is waged with city police. In such a standoff, the Northeastern cities are indeed "declining" because they have lost the ability to control what were once neighborhoods of profit and social order.

THE RISE OF THE SUNBELT CITIES AND THE "YANKEE IMMIGRANT"

Since 1950, while most of the older Northeastern cities were suffering a relentless loss in their white, skilled, and upwardly mobile population, the cities of the Sunbelt were experiencing, almost without exception, significant increases in population. In the space of three decades, the population of the region doubled, from 40 to 80 million. The impact of this shift was felt in both regions: dramatic increases in urban population growth in the Sunbelt were matched by significant declines in the Northeast. This shift did not occur in an economic vacuum. People are attracted by jobs and moved by the job market. Thus, by 1972, the major cities of the Northeast had lost anywhere from 14% to 18% of their 1958 employment in manufacturing, retailing, and wholesaling. The major cities of the Sunbelt had average employment gains of between 60% and 100% in these same three sectors (Perry and Watkins, 1977). Further, fully 60% of the South's industrial growth can be traced to the capture of rapidly growing high wage industries.

The migrants attracted to the jobs generated by this competitive edge registered in the Sunbelt economy represented a substantially different group from black migrants. They were, often as not, skilled rather than unskilled, urban rather than rural, and white rather than racially marginal. In the main, therefore, they were not pushed by abject poverty to the Sunbelt. They did not move to slums, they moved very quickly to suburban type housing. They did not represent useless labor power to the emerging Sunbelt centers of accumulation; they represented profitable, integratable, workers. They did not represent a gross infusion of high cost, socially disruptive, citizens, they represented consumer affluence and a burgeoning housing market. And, quite often, they were Northeasterners who were tired of high taxes, high welfare burdens, and a society shattered by the disorganized impact of large numbers of an "unreasonably" militant racial underclass.

Therefore, this last wave of urban migrants does not, for the most part, represent a massive infusion of a new low income urban underclass. The urban poverty of Southern and Southwestern cities preceded the new migrants by a century and a half (Watkins and Perry, 1977). In fact one of the attractive features for many labor intensive Northeastern industries was this large regional pool of docile, low skilled labor (Harris, 1952; Fuchs, 1963). Thus, it was not the poverty of the migrants, but rather the poverty of the region which helped stimulate the economic shift and the migration.

In as much as the growth of Sunbelt cities has been, in part, predicated on these differences, such cities have come to represent a new legitimation of the city as a center of profit rather than a center of wasteful social service delivery. As such, the "rise" of the Sunbelt cities does not represent the practice of a new urban function in America: it represents a regional reaffirmation of the prevailing urban function. Returning to the dilemma posed at the beginning of this essay, the apparent contradiction in the distinction between the "poverty-stricken Northeast" and the "low income Sunbelt" can now be resolved.

SUBEMPLOYMENT: THE NORTHEAST VERSUS THE SUNBELT

We posited in the introduction that the economic "rise" or "decline" of a city does not irrevocably alter the depressed life-style conditions of poverty. What is altered during such periods of change in productivity is the ability of the urban center to economically utilize low income, unskilled residents. Using a subemployment index as a measure of poverty in the economically productive Sunbelt and the declining Northeast, we will now test this assertion, especially in light of the previously discussed waves of urban immigration.

We use the subemployment index to indicate the economic condition of the urban poor in both the "affluent" Sunbelt and the "depressed" Northeast. Further, this index represents an attempt to overcome the limitations inherent in using the incomplete measure of unemployment. Unemployment, at best, tells us only the amount of people affected by cyclical movements in the job market. Thus, in addition to those officially designated as unemployed by the Bureau of Labor Statistics, the subemployment index includes the following components:

- *The discouraged jobless.* This includes workers who say that they want a job, but for a number of reasons are not looking. The reasons include such economic ones as lack of transportation or that jobs are not available and that looking would be a waste of time; also such personal ones as family responsibility, an absence of adequate day care facilities, or ill health which could be alleviated by adequate medical attention. Since they are not actively looking for work, the discouraged workers are reported neither as unemployed nor as members of the labor force. Yet since these people indicate a desire for work, an adequate indication of the impact of a city upon the social well-being of its poor cannot be constructed by simply relegating these residents to the status of nonpersons.

• *Involuntary part-time workers.* This category separates persons who work part time because they are unable to find full time work from persons working full time by choice. It also represents a departure from currently accepted Bureau of Labor Statistics practices which count any individual as fully employed irrespective of how many hours per week they work. But again, following the traditional guidelines would only hide an important aspect of a city's performance with respect to its employment capacity.

• *Workers earning substandard wages.* Qualitatively and quantitatively this is the most significant component of subemployment and represents the one element which has traditionally served as the watershed in measuring the potential utility of human beings in the city. The local economy can be said to be providing adequate avenues out of a poverty life-style if there is significant movement from substandard jobs to "good" jobs. While the subemployment index cannot measure the movement of individuals between these groups of jobs since time series data are not available, it can determine the percentage of workers who, at one specific moment, fall on either side of the cutoff point used to separate a good job from a bad job. Two possible standards will be used to enumerate workers who are fully employed but earning substandard wages. The Bureau of Labor Statistics has defined a lower level family budget which distinguishes a living wage from a substandard wage. The national urban average "lower level" family budget in 1970 was $6,960 per year, or about $3.50 per hour for a family of four; and in 1973 it was $8,181 per year or about $4.00 per hour. The Social Security Administration has, on the other hand, defined for 1970 $4,200 or $2.00 per hour as an adequate income for a family of four, and this is the basis of the official poverty line.

The data for this index is derived from the 1970 Census Employment Survey conducted by the Bureau of Labor Statistics in the low income neighborhoods of 51 major central cities. Extensive questionnaires were administered in each inner-city area to determine the respondent's recent work experience as well as past labor market history. In this fashion, unemployed workers were identified and separated from the discouraged jobless and those correctly listed as not in the labor force. Similarly, involuntary part-time workers were accurately distinguished from those working full time but earning less than $2.00 per hour, between $2.00 and $4.00 per hour, or greater than $4.00 per hour. Overall, we have constructed separate indices for two categories of low income residents: (1) primary wage earners of each family unit and (2) individuals 16 years or older and not in school.

TABLE 1
COMPARISON OF INNER-CITY SUBEMPLOYMENT RATES FOR INDIVIDUAL WAGE EARNERS IN 18 SELECTED NORTHEASTERN AND SUNBELT CITIES (in percentages)

	Unemployed			Underemployed				Total
Cities:	(1) Unem-ployed	(2) Discour-aged Workers	(3) Sub-total (1+2)	(4) Involun-tary Part Time Workers	(5) Full Time Workers $0.00-$2.00	(6) Full Time Workers $2.00-$3.50	(7) Sub-total (4+5+6)	(3+7)
			a. Northeastern Innercity Subemployment					
New York	5.6	30.7	36.3	1.6	6.1	26.9	33.6	69.9
Chicago	7.6	28.2	35.8	2.2	7.8	26.0	36.0	71.8
Philadelphia	7.6	12.4	20.0	4.0	10.1	25.3	39.4	59.4
Detroit	10.1	27.4	37.5	3.4	8.9	17.4	29.7	67.2
Boston	5.7	32.7	38.4	1.8	6.3	20.7	28.8	67.2
Pittsburgh	7.8	20.8	28.6	2.5	10.8	24.8	38.1	66.7
Cleveland	6.4	27.1	33.5	3.0	9.1	22.7	34.8	68.3
Newark	7.4	31.3	38.7	2.2	8.6	24.3	35.1	73.8
Buffalo	6.9	29.6	36.5	2.6	9.0	19.8	31.4	67.9
Average	7.2	26.6	33.8	2.5	8.5	23.1	34.1	67.9
Summary	10.6	39.2	49.8	3.7	12.5	34.0	50.2	100.0
			b. Sunbelt Innercity Subemployment					
Memphis	8.8	22.3	31.1	3.8	21.5	19.9	45.2	76.3
Birmingham	7.5	25.2	32.7	3.3	19.5	17.9	40.7	73.4
Oklahoma City	6.3	22.5	28.8	3.2	19.6	18.1	40.9	69.7
Miami	8.3	19.0	27.3	4.2	19.6	22.5	46.3	73.6
Fort Worth	8.6	18.6	27.2	4.0	20.6	21.0	45.6	72.8
Houston	4.7	19.8	24.5	3.9	17.2	22.8	43.9	68.4
Dallas	7.3	18.9	26.2	4.5	19.3	24.4	48.2	74.4
Phoenix	7.2	25.1	32.3	4.2	12.9	21.1	38.2	70.5
Atlanta	6.3	23.4	29.7	4.1	18.2	23.1	45.4	75.1
Average	7.2	21.7	28.9	3.9	18.7	21.2	43.8	72.7
Summary	9.9	29.8	39.7	5.4	25.7	29.2	60.2	100.0

SOURCE: U.S. Bureau of the Census, 1970 Census of Population and Housing, (1972). Census Employment Survey for Selected Low Income Neighborhoods. Washington, D.C.: U.S. Government Printing Office.

TABLE 2
COMPARISON OF INNER-CITY SUBEMPLOYMENT RATES FOR THE PRIMARY WAGE EARNER IN EACH FAMILY IN 18 SELECTED NORTHEASTERN AND SUNBELT CITIES (in percentages)

	Unemployed			Underemployed				Total
	(1)	(2)	(3)	(4)	(5)	(6)	(7)	
Cities:	Unemployed	Discouraged Workers	Subtotal (1+2)	Involuntary Part Time Workers	Full Time Workers $0.00-$2.00	Full Time Workers $2.00-$3.50	Subtotal (4+5+6)	(3+7)
a. Northeastern Innercity Subemployment								
New York	4.8	22.5	27.3	1.6	5.0	33.2	39.8	67.1
Chicago	4.8	16.9	21.7	1.7	4.8	14.8	21.3	43.0
Philadelphia	5.2	6.2	11.4	3.9	6.6	28.3	38.8	50.2
Detroit	7.6	19.6	27.2	3.4	7.2	20.9	31.5	58.7
Boston	5.0	21.8	26.8	1.6	4.6	22.7	28.9	55.7
Pittsburgh	6.6	14.4	21.0	2.6	6.8	28.5	37.9	58.9
Cleveland	4.8	12.0	16.8	2.5	5.9	23.2	31.6	48.4
Newark	6.7	23.7	30.4	1.6	5.4	29.2	36.2	66.6
Buffalo	5.5	20.9	26.4	2.2	7.0	24.8	34.0	60.4
Average	5.6	17.6	23.2	2.3	5.9	25.1	33.3	56.5
Summary	9.9	31.2	41.1	4.1	10.4	44.4	58.9	100.0
b. Sunbelt Innercity Subemployment								
Memphis	4.6	11.3	15.9	3.3	17.2	31.7	52.2	68.1
Birmingham	3.7	12.8	16.5	2.9	16.9	27.2	47.0	63.5
Oklahoma City	5.0	14.1	19.1	3.4	14.5	30.5	48.4	67.5
Miami	6.7	9.7	16.4	4.2	13.0	29.9	47.1	63.5
Fort Worth	5.2	8.5	13.7	4.0	12.1	28.5	44.6	58.3
Houston	2.5	8.0	10.5	3.7	13.0	32.3	49.0	59.5
Dallas	5.3	10.7	16.0	3.9	14.8	30.1	48.8	64.8
Phoenix	5.2	15.0	20.2	3.8	9.4	25.5	38.7	58.9
Atlanta	4.4	14.9	19.3	3.9	9.0	21.8	34.7	54.0
Average	4.7	11.6	16.3	3.7	13.3	28.6	45.6	62.0
Summary	7.6	18.7	26.3	6.0	21.4	46.1	73.5	100.0

SOURCE: U.S. Bureau of the Census, 1970 Census of Population and Housing, (1972). Census Employment Survey for Selected Low Income Neighborhoods. Washington, D.C.: U.S. Government Printing Office.

In Tables 1 and 2 both variants of the subemployment index are computed for 18 of the largest central cities in the industrial northeast and the Sunbelt. These cities can be properly regarded as prime centers of Northeastern and Sunbelt productivity. Hence, with this sample, both the quantitative and qualitative dimensions of the job-related sources of low income patterns found in the Northeast and the Sunbelt should be apparent.

As the overall subemployment totals suggest, the rising levels of Sunbelt affluence have done little for the poor. The total magnitude of poverty as captured by the two indices is almost identical, with the Sunbelt showing a slightly greater incidence of subemployment than the Northeast.

At the qualitative level, regional variations in subemployment patterns suggest that poverty represents very different urban conditions in the two regions. If we take the first two components—i.e., the unemployed and the discouraged jobless—we can derive a first major qualitative dimension of subemployment: the lack of opportunity for work of any quality. Measured along these two components, the Northeastern central cities clearly emerge as less successful than their Sunbelt counterparts. A person residing in the ghetto area of a major Northeastern central city will much more likely to be without a job—either because he is unable to locate one or more likely because he perceives that none are available and therefore searching would be futile. In fact, among the old cities, the discouraged jobless component is responsible for 49.8% of the total subemployed population. Limiting the definition of the relevant labor supply to encompass only the primary wage earner in each family widens the disparity between Northeast and Sunbelt cities although in all instances the magnitude of unemployment and discouraged jobless is less for family heads than for all individuals. Joblessness makes up 41.1% of the subemployed in Northeastern cities and 26.3% of the subemployed in Sunbelt cities.

The remaining three components of the subemployment index comprise a second major dimension of the urban labor market: substandard wages and the inability to find full-time work. On this basis, the Northeastern cities have clearly outperformed the Sunbelt irrespective of which subemployment index is used. The evidence is particularly striking for the category enumerating full-time workers earning less than $2.00 per hour. A worker, either individual or head of household, in a Sunbelt city is more than twice as likely to be working full time and receiving wages which are insufficient to provide a level of income above the poverty level. Ghetto residents in new cities, while much more likely to be working are also more likely to be earning substandard wages. In the Northeast, however, the incidence of joblessness is greater but for those individuals or family heads who are working, the labor market has provided them with a larger share of good jobs.

Thus poverty in the Sunbelt is in large part the result of low paying jobs, with 73.5% of subemployed inner-city heads of households employed at jobs which do not bring them a living wage. The pattern is less dramatic (58.9%) but similarly the case when Sunbelt subemployment is traced for individuals. In both cases, these characteristics of Sunbelt poverty far exceed the percentage of poor working at substandard jobs in the Northeast. Relative to the Sunbelt, substantially larger shares of the Northeastern urban poor are unemployed or refuse to look for jobs in a market that does not provide an avenue out of material deprivation. They see little difference between taking a job which guarantees them poverty and remaining unemployed.

What emerges from this analysis is the conclusion that people living below the poverty line in the urban South and Southwest represent a profitable resource to be exploited by employers of low skilled labor; while people living on poverty in the North are more likely to be unemployed because there is no job, or because they cannot find a job which meets their qualitative needs. As such, poverty levels in the North represent people who are economically unmanageable or unuseable; these people represent a barrier to profit and can emerge as a threat to the predominant social order.

DOMINANT VISIONS FOR THE URBAN FUTURE

To state this most recent pattern of growth another way, Sunbelt poverty is maintained and managed by the private sector, while Northeastern poverty, now economically unmaintainable, is becoming increasingly the management problem of the state. What is a "cheap labor" supply in the South, is a source of rising "fiscal crisis" in the Northeast, demanding, from the dominant perspective, a damaging *misallocation* of public funds away from surplus stimulating investment ventures and into redistributive social programs designed to control social unrest and somehow restrain or reshape the poor into some accumulatively useful group. From this perspective, therefore, urban poverty comes to be confronted, not because it represents human misery and a center of declining social well-being, but because it represents a failure in the city as a center of profit.

As such, the *primary* issue of social reform in the late 1970s has become reform of the social welfare and urban renewal structures and not of the declining conditions of human social well-being. Norton Long characterizes the shift of industries and people out of the Northeastern cities, leaving large numbers of economically useless and socially costly residents behind, as residential and industrial "disinvestment" of our historically dynamic centers of profit. For Long, rising conditions of urban blight are not viewed as failures in social well-being but as economic failures: this disinvestment manifests itself in "empty, blighted, gutted, and abandoned factories, office buildings, warehouses, lofts, houses and apartments. The reasons for this disinvestment have to be faced if the process is to be halted and reversed. The single most important explanation is profitability" (Long, 1977:51). City expenditures are being sadly "misallocated" in his eyes; they "are treated as pure 'merit consumption' and not as investments of scarce resources that, to some important degree, must generate a return if they are to be sustained" (Long, 1977:50).

As if heeding the siren call of Long, two new highly similar directions are emerging in urban policy which place primary emphasis upon the reform of what they view as *misallocative* urban programs that generate no return on their "investment." One perspective catalogues cities in terms of healthy and dying sections. It argues that urban policy should be directed toward the economically and residentially renewable areas, leaving blighted areas to die. The second approach suggests that both private and public policies must adhere to the realities of the societal crisis emerging in the Northeast and cut back on social spending, rising wage floors, costly labor benefits, inflated tax programs and other redistributive expenditures which divert scarce funds from region-renewing investment strategies. If the regional pattern of urban disinvestment is to be halted, then, from this perspective, Eastern cities cannot afford to continue strategies which are not devoutly attractive to accumulative interests. Thus, while one approach would have us discard sections of cities, or whole cities themselves, which have outlived their primary accumulative utility, the other approach would have us discard public and private activities which divert too much surplus from accumulatively attractive reinvestment schemes.

The urban death theorists supply a new analytic vocabulary for the increasing human misery and communal blight found in the cities of the Northeast. They relegate sections of our older cities, and some entire cities, to the level of economically useless backyard "sandboxes" (Sternlieb, 1971) and to the socially illegitimate status of barren Indian "reservations" (Long, 1971). William Baer (1976) takes the analysis one step further: these cities, and the poor who reside in them, are really representative of urban "cemeteries." They are actually "dying." They have outlived their profit function and like any organism whose function has waned and which can no longer adapt to change in its environment, natural decline and death are inevitable. Like latter day Social Darwinists, they assume the accumulative function to be the "natural" urban state, and progressively insurmountable failures in the accumulative power of this state must eventually come to be viewed as a sign of death. Baer infers that ineffective social renewal programs, after a certain time, are no more useful to the urban areas than embalming fluid is to a corpse—they preserve the corpse but do not generate its economic rebirth. Those who would continue to pour money and programmatic support into the most blighted our our cities are, in Baer's eyes, no more than urban morticians. Hence, our policy agenda for the urban future should include a contemplation of "the various aspects of urban death and . . . suggest what can be done to ameliorate its consequences. Contemplation is meant just that: a considered exposition that treats urban death as very much in the *natural order* of things, to be taken in stride—not an apocalyptic pronouncement in the currently fashionable vein" (Baer, 1976:4).

Such policies have already begun to emerge. In a recent article, Roger Starr (1976) argues that in order to rejuvenate New York, we must make it smaller. To do this he suggests a form of selective neighborhood *triage* whereby certain sections of the city (including, in particular, the Brownsville section described earlier by Jimmy Breslin) would be systematically allowed to self-destruct with the help and encouragement of the city and the nation. According to Starr, we must begin to think in terms of shrinkage because this is the fate which has been prescribed for New York City by the ultimate arbiter of all economic decisions—the Invisible Hand.

While this first group of policy advocates agrues that the best way to manage the crisis of poverty is to "pull the plug" on the terminally ill patient, the other approach suggests a drastic alteration in the mode of treatment. Their new prescription of social control suggests drastic reductions in the amounts and forms of redistributive medicines. The approach in essence punishes cities for their misallocative excesses and accumulative shortcomings. It calls for a substantial change in the amount and character of public and private sector activities in order to attract new industrial activity and cut down on the drain of disinvestment. In calling upon Northeastern governors to create an economy renewing Regional Energy Development Corporation, Felix Rohatyn captures the essence of this approach well. He argues that the billions of dollars such an investment could generate will only be forthcoming when and if the Northeastern states and cities are willing to cut back on public service spending. In order to attract such new investment and renewed industrial activity he has argued: "we need to make changes in the (business) tax structure, or changes are needed in union work rules. . . . Because of the possibility of economic activity, the governors will accept political change. What we are creating is a way to give politicians the excuse to do what they know ought to be done anyway. That's the way we got reforms in New York City's government. Every few weeks another bond issue had to be sold. Each required concessions" (McManus and Weil, 1976:375).

The concessions Rohatyn and his Big MAC fiscal oversight body exacted from the workers and poor of New York flow directly from the assertion that "political change" in our Northern cities must constitute a reduction in the "wasteful" excess of "misallocated," overly redistributive, policies in an urban community whose role is primarily accumulative.

Recently, Daniel Moynihan has argued that the fiscal plight of the major cities of New York is in large part the result of federal programs which forced the cities to become bloated centers of welfare for unrealistic numbers of people. In a report to President Carter, he states:

> one person in six in New York City is on welfare, and . . . [the] . . . other cities in the State have great welfare rolls. If a personal note may be allowed, as far back as 1965 I persuaded the third President before the present one that this was going to happen. To use an economist's term, the welfare system is exogenous. This is the say the influence comes from outside the system. New York State did not create the present welfare

> system—the Federal government did. If the system were changed, a great many of its symptoms might disappear, which would prove they had nothing to do with New York as such. [Moynihan, 1977:S10829]

Moynihan finds it unrealistic to expect the cities of his region to supply the amount of social services they now deliver. In his mind, a reduction in such redistributive activity is essential if cities are to survive the chaotic "excesses" that continue to materialize as citizens demand more and more services from the state.

Indeed, the demands of the poor for high paying jobs and the comcomitant leisure-oriented luxuries of a consumer society, are, in Norton Long's view, also unrealistic goals for the urban poor in America (Long, 1977:55). Citing data which suggest that the primary way a poor family can expect to receive an adequate income is by having multiple earners in the family working at low wages, Long argues we must cut back on "union and liberal policy" (1977:55) which drives up the wage costs and the expectations of jobholders in low skill industries to the point where the jobs disappear altogether or the economic activity is removed to the low wage Sunbelt.

These forms of policy redirection have their roots in the overall condition of uneven urban development between the Northeast and the Sunbelt. The impact of the policies offered by Rohatyn, Long, Starr, and others would cheapen Northeastern labor, reduce corporate taxes decrease the public service burdens and otherwise ameliorate the rising social well-being costs which dulled the competitive edge of the Northeastern urban centers. Thus, just as the present day profit emanating from the Sunbelt economy now looks very much like the latter day profit levels of the Northeast, social welfare policies are now being abandoned in the Northeast in an attempt to bring industrial rim redistributive activities down to the low level of the Sunbelt (Harris, 1952).

THE DECLINE OF THE SUNBELT AND THE RISE OF URBAN SOCIAL WELL-BEING

Against all this, the "rise of the Sunbelt cities" represents a geographical shift in the "successful" practice of the city as the dominant unit of profit. The migrants to the Sunbelt, while substantially different from the other waves, have been profitably integrated into the accumulation process and are highly manageable. Therefore, at present, social reform in the South is in no danger of "erring" through a misallocation of surplus. The poor in the South are also highly employable and a substantial share of their poverty results from the low wages they receive. From this perspective, the new cities of the Sunbelt represent nothing new for the poor—their social well-being is no more ensured in Sunbelt cities than it is in cities of the Northeast.

For us, therefore, the "rise" of the Sunbelt means a new legitimation of the capitalist city and the "decline" of the Northeast is evidence of a crisis in the legitimacy of America's political economy. The new proposals of established policy analysts reflect a concern with this regionally based legitimation crisis.

If the rise of the Sunbelt and the present prescriptions for urban change reflect no alteration in the historical tradition of a permanent urban underclass, and if history repeats itself, then the time will come when Sunbelt cities begin to experience an eclipse in their social legitimacy. The advantageous "low income pool" of Southern workers will be transformed into "poverty stricken" urbanites. Indeed the cities of the Sunbelt will have achieved a state of "crisis" parity with the Northeast. As the rising social and economic tensions of the Northeastern cities have demonstrated, the poor do not go away. They do not lose their desire for human renewal; they do not lose their frustration, discouragement, and anger. While the capitalist system can temporarily escape such social dislocation by moving to the suburbs and to the Sunbelt—trading in old urban centers, like used cars, for new models—it cannot proceed in this manner forever. Unlike used cars, declining centers of profit cannot be removed to a junk yard, slipped into a coffin, or pushed to a controlled reservation. They remain filled with humans who are more than simply units of labor, more than "good business climates."

The social unrest of a declining Northeast is a sign of a rising crisis in the legitimacy of the city as first and foremost a center of profit and it cannot be hidden by the "rise" of the Sunbelt. More than ever, it is time to recast the definition of the city in America. It is time to measure urban success in terms of the climate of social well-being the cities represent for people. We must become aware of economic measures of urban health that have at their source the economic exploitation of human lives and temper our definitions of urban growth accordingly. We should test social welfare programs against criteria of social well-being rather than social control. Unless we begin this task of political economic renewal, and consider the American city first and foremost as a center for people rather than profit, then the traditional definitions of the "rise" and "decline" of American cities will become meaningless measures of American development for more than simply the urban poor.

REFERENCES

ADDAMS, J. (1910). Twenty years at Hull House. New York: Macmillan.

BAER, W.C. (1976). "On the death of cities." Public Interest, 45(fall): 3-19.

BOOKCHIN, M. (1973). The limits of the city. New York: Harper and Row.

BRACE, C.L. (1872). The dangerous classes of New York and twenty years work among them. New York: Wynkoop and Hallenbeck.

BREMNER, R.H. (1956). From the depths: The discovery of poverty in the United States. New York: New York University Press.

BRESLIN, J. (1971). "Moonwalk on Sutter Avenue." Pp. 231-234 in D.M. Gordon (ed.), Problems in political economy: An urban perspective. Lexington, Mass.: D.C. Heath.

CHUDACOFF, H.P. (1975). The evaluation of American urban society. Englewood Cliffs, N.J.: Prentice-Hall.

CLOWARD, R.A., and PIVEN, F.F. (1974). The politics of turmoil: Essays on poverty, race and the urban crisis. New York: Random House.

DICKENS, C. (1972). American notes for general circulation. J.S. Whitley and A. Goldman (eds.). Baltimore, Md.: Penguin.

FARLEY, R. (1977). "Trends in racial inequalities: Have the gains of the 1960s disappeared in the 1970s? American Sociological Review, 42(2): 189-208.

FEAGIN, J.R. (forthcoming). Race and ethnic relations. Englewood Cliffs, N.J.: Prentice-Hall.

FUCHS, V.R. (1963). Changes in the location of manufacturing in the United States since 1929. New Haven, Conn.: Yale University Press.

GAMBINO, R. (1975). Blood of my blood: The dilemma of the Italian-American. Garden City, N.Y.: Doubleday.

GINGER, R. (1973). Altgeld's America: The Lincoln ideal versus changing realities. New York: New Viewpoints.

GREELEY, A.M. (1972). That most distressful nation: The taming of the American Irish. Chicago: Quadrangle.

GREELEY, H. (1864). "Tenement houses–their wrongs." New York Daily Tribune, November 23:4.

GRISCOM, J.H. (1945). The sanitary condition of the laboring population of New York and suggestions for its improvement. New York: Harper and Row.

GROB, G.N. (1969). Workers and utopia: A study of ideological conflict in the American labor movement, 1865-1900. Chicago: Quadrangle.

HARRIS, S.E. (1952). The economics of New England: Case study of an older area. Cambridge: Harvard University Press.

HARTLEY, R. (1842). "An historical, scientific and practical essay on milk, as an article of human sustenance; with a consideration of the effects consequent upon the present unnatural methods of producing it for the supply of large cities." Eighth annual report of the New York Association for Improving the Condition of the Poor. New York: Jonathan Leavitt.

JAMES, D.B. (1972). Poverty, politics and change. Englewood Cliffs, N.J.: Prentice-Hall.

KATZNELSON, R. (1976). "The crisis of the capitalist city: Urban politics and social control." Pp. 214-229 in W.D. Hawley et al. (eds.), Theoretical perspectives on urban politics. Englewood Cliffs, N.J.: Prentice-Hall.

LONG, N.E. (1971). "The city as reservation." Public Interest, 25(fall):22-38.

--- (1977). "A Marshall plan for cities?" Public Interest, 46(winter):49-60.

LuBOVE, R. (1962). The progressives and the slums: Tenement house reform in New York City, 1890-1917. Pittsburgh: University of Pittsburgh Press.

McCABE, J.D. (1882). New York City by sunlight and gaslight, a work descriptive of the great metropolis. Philadelphia: National Publishing Company.

McMANUS, M.J., and WEIL, F.A. (1976). "No one is in charge." Empire State Report, (October-November):364-375.

MORSE, S.F.B. (1835). Foreign conspiracy against the liberties of the United States. New York: Leavitt and Lord.

MOYNIHAN, D.P. (1977). The federal government and the economy of New York State. Congressional Record–Senate. June 27, pp. S10829-S10833.

PERRY, D.C., and WATKINS, A.J. (1977). "To kill a city: A critical reevaluation of the status of yankee and cowboy cities." Studies in Politics: Series I: Studies in Urban Political Economy, Paper no. 6. Austin: University of Texas.

PIVEN, F.F. (1974). "The great society as political strategy." Pp. 271-283 in R.A. Cloward and F.F. Piven, The politics of turmoil: Essays on poverty, race and the urban crisis. New York: Random House.

PIVEN, F.F., and CLOWARD, R.A. (1971). Regulating the poor: The functions of public welfare. New York: Random House.

Report of the National Advisory Commission on Civil Disorder (1968). New York: Random House.

STARR, R. (1976). "Making New York smaller." New York Times Magazine, November 1:32-33, 99-106.

STERNLIEB, G. (1971). "The city as sandbox." Public Interest, 25(fall):14-21.

TRATTNER, W.I. (1974). From poor law to welfare state: A history of social welfare in the United States. New York: Free Press.

U.S. Bureau of the Census (1972). 1970 Census of Population and Housing, Census employment survey for selected low income neighborhoods. Washington, D.C.: U.S. Government Printing Office.

WADE, R.C. (1959). The urban frontier: The rise of western cities, 1790-1830. Cambridge: Harvard University Press.

WARNER, S.B., Jr. (1968). The private city: Philadelphia in three periods of its growth. Philadelphia: University of Philadelphia Press.

WATKINS, A.J., and PERRY, D.C. (1978). "Regional change and uneven urban development." In D.C. Perry and A.J. Watkins (eds.), The rise of the sunbelt cities. Beverly Hills, Calif.: Sage.

THE AUTHORS

WILLIAM D. ANGEL, Jr. is a Ph.D. candidate in the Department of Government at the University of Texas at Austin. He is presently completing a dissertation on the role of entrepreneurship in urban America and has published work in this area and elsewhere.

MURRAY BOOKCHIN is a Professor in the Institute for Metropolitan Studies at Ramapo College of New Jersey. He has written *Post-Scarcity Anarchism, The Limits of the City*, and a forthcoming work entitled *Urbanization Without Cities.*

GURNEY BRECKENFELD is a member of the Board of Editors of Fortune Magazine, a veteran journalist, author and authority on building, housing and urban affairs. For the past 26 years he has served on the staff of four Time Inc. magazines and is the coauthor of the book, *The Human Side of Urban Renewal* and the author of *Columbia and the New Cities.*

GENE BURD is an Associate Professor of Urban Journalism at the University of Texas at Austin. He was one of the last residents of Chicago's original Hull-House, studied "Magazines and the Metropolis" on a grant from the Magazine Publishers Association of New York, and served as a consultant for the Twin Cities Metropolitan Council in Minneapolis–St. Paul.

ROBERT B. COHEN is currently serving as the Transnational Corporation Affairs Officer at the U.N. Center on Transnational Corporations. He is the author of a forthcoming book entitled *The Modern Corporation and the Advanced Corporate Services Complex* and is presently editing a work by the late Stephen Hymer entitled *The Multinational Corporation: A Radical Critique.*

ROBERT E. FIRESTINE is an Associate Professor of Economics and Political Economy at the University of Texas at Dallas. He is coauthor of a forthcoming book, *Regional Growth and Decline in the United States: The Rise of the Sunbelt and the Decline of the Industrial Northeast.*

ARNOLD FLEISCHMANN received a B.A. in economics in 1973 from the College of St. Thomas, St. Paul, Minnesota. Prior to writing his article on San Antonio boosterism, he lived in San Antonio for several years. Currently he is serving on the staff of Congressman Bob Eckhardt.

DAVID M. GORDON teaches economics at the Graduate Faculty of the New School for Social Research in New York City. Most recently he has written *Theories of Poverty and Underemployment* and edited *Problems in Political Economy: An Urban Perspective.*

PETER A. LUPSHA is an Associate Professor of Political Science at the University of New Mexico. He is the author of a wide ranging set of studies and essays on political movements, urban violence, and domestic public policy.

DAVID C. PERRY is an Associate Professor of Government at the University of Texas at Austin. He is the author of *Police in the Metropolis* and the coeditor of *Violence as Politics.* At present he is coauthoring *The Political Economy of Regional Change* with Alfred Watkins.

WALT W. ROSTOW is a Professor of Economics and History at the University of Texas at Austin. He has authored numerous articles and books, including *The Stages of Economic Growth* and a forthcoming three volume work entitled *The World Economy: History and Prospect.*

WILLIAM J. SIEMBIEDA is an Associate Professor of Planning at the University of New Mexico and Senior Research Associate at the Albuquerque Urban Observatory. He is the author of a series of studies on the impact of urban and national public policy.

ALFRED J. WATKINS is an Instructor in the Department of Government at the University of Texas at Austin. In 1977 he received his Ph.D. from the Department of Economics of the Graduate Faculty of the New School for Social Research. He has written several articles and is presently coauthoring the book *The Political Economy of Regional Change,* with David C. Perry.

NOTES

NOTES

NOTES

NOTES

NOTES

NOTES

NOTES

NOTES

NOTES

NOTES